# Semiconductor Cleanrooms Management Essentials

# Preface

Welcome to the comprehensive guide on managing a cleanroom environment. Whether you are new to cleanroom operations or seeking to enhance your existing knowledge, this book is designed to provide you with a thorough understanding of the principles, protocols, and practices essential for maintaining a pristine and controlled environment.

In today's high-tech industries such as semiconductor manufacturing, biotechnology, and pharmaceuticals, cleanrooms play a crucial role in ensuring product quality, process integrity, and safety. However, managing a cleanroom effectively requires more than just maintaining cleanliness; it demands a deep understanding of various aspects ranging from safety procedures to equipment usage, contamination control to documentation practices.

This book is structured to serve as your comprehensive reference, covering a wide array of topics essential for cleanroom management. From safety orientations to gowning procedures, from contamination control to waste disposal protocols, each chapter delves into specific aspects of cleanroom management, providing practical insights, best practices, and actionable guidelines.

Throughout these pages, you will find discussions on critical subjects such as cleanroom etiquette, tool and equipment usage, environmental monitoring, and documentation practices. Moreover, we address specialized topics including cleanroom classification and standards, access control, and environmental monitoring, among others.

Furthermore, this book goes beyond the fundamental operational aspects of cleanrooms. It explores the role of a cleanroom manager in coordinating facility modifications, resolving equipment issues, and collaborating with scientific projects. It also delves into the maintenance procedures necessary for ensuring optimal cleanroom performance and offers guidance on inventory management and communication strategies.

Whether you are a cleanroom manager, technician, researcher, or industry professional, this book aims to equip you with the knowledge and skills necessary for navigating the complexities of cleanroom management. It is our sincere hope that this resource serves as a valuable companion in your journey towards maintaining the highest standards of cleanliness, safety, and efficiency within your cleanroom environment.

Thank you for embarking on this educational journey with us.

Sincerely,

Dr. Aboulwafa Singer

## Acknowledgement

I would like to express my deepest gratitude to Professor Salah Obayya, my Ph.D. supervisor, for his unwavering support and invaluable guidance throughout my academic and professional journey. His encouragement, insight, and wisdom have been pivotal, not only in shaping my research but also in my broader approach to science and engineering.

During my Ph.D. studies, Professor Obayya's constant motivation made life easier and more productive for me. I recall a conversation we had when I mentioned my aspiration to write a comprehensive book on the fundamentals of Telecommunication Engineering. My goal was to create a resource that covered a broad range of concepts, from mathematics and statistics to Fourier transforms. His words of advice left a lasting impression on me. He highlighted the importance of making complex concepts accessible and manageable for readers, emphasizing that a book with such wide-ranging topics could become overwhelming. This advice was transformative, and it influenced my approach when writing this book, *Semiconductor Cleanrooms Management Essentials*.

Professor Obayya himself is a distinguished scholar and author, having authored several seminal works in photonics and computational electromagnetics. His academic contributions and leadership as Director of the Centre for Photonics and Smart Materials at Zewail City of Science and Technology, as well as his roles at universities around the world, have been a continual source of inspiration for me. His ability to distill complex scientific principles into practical, digestible knowledge is something I aspire to emulate.

This book would not have been possible without Professor Obayya's mentorship, and for that, I am forever thankful.

With sincere gratitude,

Dr. Aboulwafa Singer

# Contents

# Chapter 1: Introduction to Cleanroom Management

## 1.1 Understanding the Importance of Cleanroom Environment

### 1.1.1 Importance of cleanrooms in Semiconductor Industries

The early stages of semiconductor development marked a transformative period in technology, paving the way for the modern electronics industry. In the mid-20th century, pioneers such as William Shockley, John Bardeen, and Walter Brattain ushered in the era of semiconductor devices with the invention of the transistor at Bell Labs in 1947. This breakthrough laid the foundation for a myriad of electronic applications, ranging from radios and televisions to computers and telecommunications systems.

However, the nascent semiconductor industry faced significant challenges in achieving reliable and reproducible device performance. One of the most formidable obstacles was contamination. In the quest to fabricate increasingly miniaturized and complex semiconductor components, even the tiniest impurities could wreak havoc on device functionality.

Contamination could arise from various sources, including airborne particles, dust, moisture, and chemical residues. These contaminants could interfere with the delicate semiconductor fabrication processes, leading to defects in the final devices. In the early days, semiconductor manufacturing facilities lacked the stringent cleanliness standards and controlled environments that are commonplace today. As a result, contamination-related yield losses were a frequent occurrence, impeding the scalability and commercial viability of semiconductor technologies.

Moreover, the sensitivity of semiconductor materials to contamination posed formidable technical challenges for researchers and engineers. Semiconductor devices rely on precise doping and patterning processes to manipulate the electrical properties of materials such as silicon. Even trace amounts of contaminants could alter the conductivity, carrier mobility, and other critical parameters of semiconductor materials, compromising device performance and reliability.

To address these challenges, early semiconductor pioneers experimented with various cleanliness measures and fabrication techniques to minimize contamination. Cleanrooms, although rudimentary by modern standards, emerged as essential environments for semiconductor fabrication, offering controlled conditions free from airborne pollutants and particulates. Additionally, advancements in materials purification, deposition methods, and lithography techniques played a crucial role in enhancing device quality and yield.

Despite these early efforts, contamination remained a persistent issue in semiconductor manufacturing throughout the mid-20th century. It wasn't until the advent of stringent cleanroom standards and advanced contamination control technologies in the 1960s and 1970s that the semiconductor industry made significant strides in overcoming contamination-related challenges. These developments paved the way for the mass production of reliable and high-performance semiconductor devices, fueling the rapid growth and technological innovation that characterize the modern electronics landscape.

**In summary,** the early stages of semiconductor development were characterized by groundbreaking innovations and persistent challenges. Contamination posed a formidable barrier to progress, requiring concerted efforts and technological advancements to overcome. The evolution of cleanroom technology and contamination control measures played a pivotal role in shaping the trajectory of the semiconductor industry, laying the groundwork for its transformation into a cornerstone of the global economy.

## 1.1.2 Importance of cleanrooms in Pharmaceutical Industries

The pharmaceutical industry plays a vital role in safeguarding public health by developing and manufacturing life-saving drugs and therapies. Central to this mission is ensuring the safety, efficacy, and quality of pharmaceutical products, a task that hinges critically on maintaining strict cleanliness standards throughout the manufacturing process. Cleanrooms, with their controlled environments and stringent contamination control measures, are indispensable facilities in pharmaceutical manufacturing, serving as the cornerstone of drug production and quality assurance.

In the early days of pharmaceutical manufacturing, contamination was a pervasive and insidious threat to product quality. The presence of airborne particles, microorganisms, and other contaminants could compromise the integrity of drug formulations, leading to reduced efficacy, safety concerns, and regulatory non-compliance. Moreover, the sensitive nature of many pharmaceutical compounds necessitated precise control over environmental conditions to prevent degradation and ensure stability.

Cleanrooms emerged as a solution to these challenges, providing pharmaceutical manufacturers with controlled environments where contamination levels are meticulously regulated and monitored. These environments minimize the risk of product contamination during critical manufacturing processes such as compounding, filling, and packaging. By maintaining low particle counts, controlled temperature and humidity levels, and stringent sanitation protocols, cleanrooms help safeguard the purity and potency of pharmaceutical products from raw material preparation to final packaging.

The importance of cleanrooms in pharmaceutical manufacturing extends beyond product quality to regulatory compliance and consumer safety. Regulatory agencies such as the U.S. Food and Drug Administration (FDA) and the European Medicines Agency (EMA) impose stringent cleanliness standards on pharmaceutical manufacturers to ensure the safety, efficacy, and consistency of drug products. Compliance with these standards is non-negotiable for obtaining regulatory approvals and market authorization, making cleanrooms an essential component of pharmaceutical quality systems.

Contamination-related issues in pharmaceutical manufacturing can have far-reaching consequences, ranging from costly product recalls to reputational damage and legal liabilities. Cleanrooms mitigate these risks by providing a controlled environment where contamination risks are minimized, thereby enhancing product quality, reliability, and safety. Furthermore, adherence to strict cleanliness standards instills confidence among healthcare professionals, regulatory authorities, and patients in the quality and integrity of pharmaceutical products.

As the pharmaceutical industry continues to innovate and diversify, cleanrooms remain at the forefront of efforts to ensure the quality and safety of drug products. Advances in cleanroom technology, contamination control strategies, and process optimization techniques continue to drive improvements in pharmaceutical manufacturing efficiency, productivity, and regulatory compliance. By investing in state-of-the-art cleanroom facilities and implementing best practices in contamination control, pharmaceutical manufacturers can uphold the highest standards of product quality and patient care, reaffirming their commitment to public health and well-being.

In summary, cleanrooms are indispensable facilities in pharmaceutical manufacturing, playing a crucial role in safeguarding product quality, regulatory compliance, and consumer safety. By providing controlled environments free from contaminants, cleanrooms enable pharmaceutical manufacturers to produce high-quality drugs and therapies that meet the stringent requirements of regulatory agencies and healthcare professionals. As the pharmaceutical industry evolves, cleanrooms will remain essential tools in the quest for innovation, excellence, and patient-centric healthcare.

## 1.2 Overview of Cleanroom Operations

Cleanroom operations encompass a diverse array of activities and procedures aimed at maintaining the cleanliness, integrity, and functionality of controlled environments essential for sensitive manufacturing processes. In this section, we provide an accessible overview of the key components and practices involved in cleanroom operations, shedding light on the critical role they play in various industries, including semiconductor manufacturing, pharmaceuticals, biotechnology, and healthcare.

### 1.2.1: Purpose of Cleanrooms

Cleanrooms serve as specialized facilities designed to minimize and control airborne particles, contaminants, and environmental variables that could compromise the quality, reliability, and safety of products manufactured within them. By establishing stringent cleanliness standards and implementing comprehensive contamination control measures, cleanrooms provide a controlled environment where critical processes such as semiconductor fabrication, pharmaceutical compounding, and biotechnology research can be conducted with precision and consistency.

### 1.2.2: Cleanroom Classification

Cleanrooms are classified according to their cleanliness levels, which are determined based on the concentration of airborne particles per cubic meter of air. Classification standards, such as ISO 14644 and Federal Standard 209E, specify particle count limits for different cleanroom classes, ranging from ISO Class 1 (the cleanest) to ISO Class 9 (the least clean). These classifications help ensure uniformity and consistency in cleanroom design, operation, and performance across industries and applications.

### 1.2.3: Design and Construction

The design and construction of cleanrooms are meticulously planned and executed to meet the stringent cleanliness requirements of their intended applications. Cleanroom design considerations include airflow patterns, filtration systems, material selection, surface finishes, and layout optimization to minimize contamination risks and facilitate efficient operation. Specialized construction materials and sealing techniques are employed to create airtight enclosures that prevent external contaminants from infiltrating the cleanroom environment.

### 1.2.4: Environmental Controls

Environmental controls play a crucial role in maintaining the cleanliness and integrity of cleanroom environments. These controls encompass temperature, humidity, air pressure differentials, and airflow rates, which are carefully monitored and regulated to ensure optimal conditions for manufacturing processes and personnel comfort. Advanced HVAC (heating, ventilation, and air conditioning) systems, filtration technologies, and monitoring devices are employed to achieve and maintain the desired environmental parameters within cleanrooms.

### 1.2.5: Personnel Practices

Personnel practices are essential for minimizing human-induced contamination in cleanroom environments. Strict gowning procedures, including the use of cleanroom attire such as coveralls, gloves, masks, and shoe covers, help prevent the introduction of contaminants by personnel. Additionally, training programs, hygiene protocols, and behavioral guidelines are implemented to promote awareness of cleanroom best practices and foster a culture of cleanliness among cleanroom personnel.

### 1.2.6: Cleaning and Maintenance

Regular cleaning and maintenance are imperative for preserving the cleanliness and functionality of cleanroom environments. Cleaning protocols, including surface disinfection, equipment decontamination, and floor maintenance, are conducted according to predefined schedules using specialized cleaning agents and techniques. Preventive maintenance measures, such as filter replacement, equipment calibration, and facility inspections, are implemented to identify and address potential sources of contamination or system failures proactively.

### 1.2.7: Quality Assurance and Compliance

Quality assurance and compliance are integral aspects of cleanroom operations, ensuring that manufacturing processes meet regulatory requirements and industry standards. Quality control measures, including environmental monitoring, particle counting, and microbial testing, are performed to validate cleanroom performance and product quality. Compliance with regulatory guidelines, such as Good Manufacturing Practice (GMP) and ISO standards, is essential for maintaining product integrity, regulatory approval, and customer confidence in cleanroom operations.

### 1.2.8: Continuous Improvement

Continuous improvement is a fundamental principle of cleanroom operations, driven by ongoing efforts to enhance efficiency, effectiveness, and innovation. Performance monitoring, data analysis, and feedback mechanisms are employed to identify areas for improvement and implement corrective actions. By embracing a culture of continuous improvement, cleanroom facilities can adapt to evolving technological advancements, regulatory requirements, and industry trends while striving for excellence in cleanliness, quality, and operational excellence.

**In summary,** cleanroom operations encompass a comprehensive suite of practices and procedures aimed at maintaining the cleanliness, integrity, and functionality of controlled environments essential for sensitive manufacturing processes. By adhering to stringent cleanliness standards, implementing robust contamination control measures, and fostering a culture of excellence and continuous improvement, cleanroom facilities play a vital role in ensuring product quality, regulatory compliance, and customer satisfaction across diverse industries and applications.

## 1.3 Role of a Cleanroom Manager

In the dynamic and complex environment of cleanroom operations, the role of a cleanroom manager is indispensable. Cleanroom managers are tasked with overseeing all aspects of cleanroom facilities, from planning and design to daily operations and maintenance. In this section, we delve into the multifaceted responsibilities and key competencies required of cleanroom managers, highlighting their pivotal role in ensuring the integrity, efficiency, and regulatory compliance of cleanroom operations.

### 1.3.1: Leadership and Strategic Planning

At the core of the cleanroom manager's role is leadership and strategic planning. Cleanroom managers are responsible for developing and implementing strategic plans, policies, and procedures to support the goals and objectives of cleanroom facilities. They provide vision, direction, and guidance to cleanroom staff, fostering a culture of excellence, collaboration, and continuous improvement.

### 1.3.2: Facility Management

Cleanroom managers oversee all aspects of cleanroom facility management, including design, construction, and commissioning. They collaborate with architects, engineers, and contractors to ensure that cleanroom facilities are designed and constructed to meet the specific needs and requirements of their intended applications. Cleanroom managers are also responsible for optimizing facility layout, workflow, and resource allocation to maximize efficiency and productivity.

### 1.3.3: Operational Oversight

Cleanroom managers play a crucial role in the day-to-day operations of cleanroom facilities. They develop and enforce standard operating procedures (SOPs), protocols, and guidelines to ensure consistent and compliant operation of cleanroom environments. Cleanroom managers monitor key performance indicators (KPIs), such as cleanliness levels, environmental parameters, and equipment reliability, to assess operational performance and identify areas for improvement.

### 1.3.4: Regulatory Compliance

Ensuring regulatory compliance is a top priority for cleanroom managers. They stay abreast of relevant regulations, standards, and guidelines governing cleanroom operations, such as Good Manufacturing Practice (GMP), ISO standards, and industry-specific requirements. Cleanroom managers develop and implement compliance programs, training initiatives, and audit processes to ensure that cleanroom facilities adhere to regulatory requirements and maintain certification.

### 1.3.5: Personnel Management and Training

Cleanroom managers are responsible for recruiting, training, and managing cleanroom personnel. They oversee staffing levels, job assignments, and performance evaluations to ensure that cleanroom facilities are adequately staffed and that personnel are properly trained and qualified to perform their duties. Cleanroom managers promote a culture of safety, professionalism, and continuous learning among cleanroom staff, emphasizing the importance of adherence to cleanliness protocols and best practices.

### 1.3.6: Budgeting and Resource Management

Effective budgeting and resource management are essential aspects of the cleanroom manager's role. Cleanroom managers develop and manage budgets for cleanroom facilities, equipment purchases, and maintenance activities, ensuring that financial resources are allocated efficiently and in accordance with organizational priorities. They identify cost-saving opportunities, negotiate contracts with vendors and suppliers, and monitor expenditures to optimize resource utilization and control costs.

### 1.3.7: Continuous Improvement and Innovation

Cleanroom managers are champions of continuous improvement and innovation in cleanroom operations. They foster a culture of innovation, creativity, and problem-solving among cleanroom staff, encouraging the development and implementation of novel technologies, processes, and best practices to enhance cleanliness, efficiency, and productivity. Cleanroom managers leverage feedback from stakeholders, performance data, and industry trends to drive continuous improvement initiatives and stay at the forefront of cleanroom technology and innovation.

### 1.3.8: Communication and Stakeholder Engagement

Effective communication and stakeholder engagement are critical skills for cleanroom managers. They liaise with internal and external stakeholders, including senior management, regulatory agencies, suppliers, and cleanroom users, to ensure alignment of objectives, expectations, and priorities. Cleanroom managers facilitate open dialogue, collaboration, and transparency, addressing concerns, resolving conflicts, and fostering positive relationships to support the success of cleanroom operations.

**In summary,** the role of a cleanroom manager is multifaceted and dynamic, encompassing leadership, strategic planning, facility management, operational oversight, regulatory compliance, personnel management, budgeting, continuous improvement, and stakeholder engagement. Cleanroom managers are instrumental in ensuring the integrity, efficiency, and regulatory compliance of cleanroom operations, playing a pivotal role in driving excellence, innovation, and success in cleanroom facilities across diverse industries and applications.

# Chapter 2: Safety Orientation

## 2.1 Emergency Procedures

### 2.1.1 Evacuation Plan

A cleanroom evacuation plan is a crucial component of ensuring the safety of personnel in the event of an emergency. Here's a brief overview of the key elements of a cleanroom evacuation plan:

1. **Alarm Activation:**

   - In case of an emergency, alarms such as fire alarms or gas detection alarms will be activated.

   - Personnel should be familiar with the different alarm sounds and meanings.

2. **Emergency Notification:**

   - A clear communication system should be in place to notify all personnel about the emergency.

   - This can include intercom announcements, visual signals, or automated messaging systems.

3. **Evacuation Routes:**

   - Clearly marked evacuation routes should be identified and known to all personnel.

   - These routes should lead to safe assembly points outside the cleanroom.

4. **Assembly Points:**

   - Designated assembly points outside the cleanroom provide a safe location for evacuated personnel.

   - Roll call or check-in procedures may be implemented at assembly points.

5. **Emergency Equipment:**

   - Personnel should be aware of the location of emergency equipment such as fire extinguishers, emergency exits, and first aid kits.

   - Training on the proper use of emergency equipment may be provided.

6. **Evacuation Procedures:**

- Personnel should follow established evacuation procedures, including leaving personal belongings behind and not using elevators during evacuation.

- Emergency responders may assist in guiding the evacuation.

7. **Accounting for Personnel:**

- A system for accounting for all personnel during and after evacuation is essential.

- This may involve designated individuals responsible for checking in personnel at assembly points.

8. **Emergency Services Coordination:**

- The cleanroom evacuation plan should include coordination with local emergency services.

- Emergency responders should be informed about the nature of the cleanroom and any potential hazards.

9. **Training and Drills:**

- Regular training sessions and evacuation drills should be conducted to familiarize personnel with the procedures.

- Drills help ensure a swift and organized evacuation in real emergencies.

10. **Post-Evacuation Actions:**

- After evacuation, personnel should follow further instructions from emergency responders or designated personnel.

- Debriefings may be conducted to evaluate the effectiveness of the evacuation and identify areas for improvement.

## 2.1.2 Alarm System

In a cleanroom, various alarms are implemented to ensure the controlled environment is maintained. Here are common alarms and their meanings:

1. **Temperature Alarm:**

- Meaning: Indicates a deviation from the specified temperature range.

- Mitigation: Investigate and adjust the environmental controls to bring the temperature back within the acceptable range.

2. **Humidity Alarm:**

   - Meaning: Indicates a deviation from the specified humidity range.

   - Mitigation: Investigate and adjust the environmental controls to restore humidity levels within the acceptable range.

3. **Pressure Differentials Alarm:**

   - Meaning: Indicates an imbalance in pressure differentials between cleanroom zones.

   - Mitigation: Adjust airflow systems to restore proper pressure differentials and prevent contaminants from entering the cleanroom.

4. **Airborne Particle Count Alarm:**

   - Meaning: Indicates an elevated level of airborne particles.

   - Mitigation: Investigate potential sources of particle generation, check filters, and take corrective actions to reduce particle levels.

5. **Vibration Alarm:**

   - Meaning: Indicates excessive vibrations that may affect sensitive processes or equipment.

   - Mitigation: Identify and address the source of vibrations, which may include adjusting equipment placement or isolating vibrations.

6. **Gas and Chemical Sensors Alarm:**

   - Meaning: Indicates the presence of harmful gases or chemicals beyond the acceptable limits.

   - Mitigation: Evacuate personnel, address the source of the gas or chemical, and ensure proper ventilation.

7. **Fire and Smoke Detection Alarm:**

   - Meaning: Indicates the presence of fire or smoke.

   - Mitigation: Evacuate personnel, isolate affected areas, and use fire suppression systems as needed.

8. **Emergency Power Failure Alarm:**

- Meaning: Indicates a failure in the primary power source.

    - Mitigation: Activate emergency power systems to maintain critical operations and initiate procedures for a controlled shutdown if necessary.

9. **Access Control Alarm:**

    - Meaning: Indicates unauthorized access to the cleanroom.

    - Mitigation: Investigate the breach, restrict access, and address security protocols.

10. **Equipment Malfunction Alarms:**

    - Meaning: Alerts about malfunctions in critical equipment.

    - Mitigation: Investigate the cause, perform necessary repairs, and ensure backup systems are in place.

It is important to note that specific cleanrooms may have additional alarms based on the nature of their processes and the level of control required. The mitigation actions should be part of the standard operating procedures for the cleanroom.

## 2.1.3: Emergency Exits and Their Importance

Emergency exits are critical components of any cleanroom evacuation plan, providing a means of egress in the event of an emergency such as a fire, chemical spill, or other hazardous situation. In this section, we explore the significance of emergency exits in ensuring the safety and well-being of cleanroom personnel and the importance of familiarizing oneself with their location and use.

**Importance of Emergency Exits:**

1. Rapid Evacuation: During emergencies, such as fires or chemical releases, every second counts. Emergency exits offer the fastest and safest route for personnel to evacuate the cleanroom and reach a designated assembly area or safe zone. Knowing the location of emergency exits beforehand enables individuals to evacuate quickly and efficiently, minimizing the risk of injury or harm.

2. Alternative Escape Routes: In the event that primary exit routes are blocked or inaccessible due to smoke, fire, or debris, alternative emergency exits provide additional escape routes for personnel to evacuate safely. Familiarizing oneself with the locations of multiple emergency exits allows individuals to choose the nearest and safest route based on the nature and severity of the emergency.

3. Compliance with Regulations: Regulatory agencies, such as Occupational Safety and Health Administration (OSHA) and local building codes, mandate the provision of adequate emergency exits in workplaces, including cleanroom facilities. Compliance with these regulations is essential for ensuring the safety and well-being of cleanroom personnel and avoiding potential legal liabilities in the event of an emergency.

4. Minimization of Panic: In stressful and chaotic situations, such as emergencies, clear signage and visible markings indicating the location of emergency exits help minimize panic and confusion among cleanroom personnel. By proactively identifying and familiarizing themselves with the nearest emergency exits, individuals can maintain a sense of calm and confidence, facilitating orderly evacuation procedures.

5. Facilitation of Emergency Response: Emergency exits not only enable personnel to evacuate safely but also facilitate the ingress of emergency responders, such as firefighters, paramedics, and hazardous materials teams. Clear and unobstructed emergency exits allow responders to access the cleanroom quickly and efficiently, enabling them to assess the situation, render assistance, and mitigate the emergency effectively.

**Recommendations for Personnel:**

1. **Familiarization:**

   Take the time to familiarize yourself with the locations of emergency exits in the cleanroom facility. Pay attention to signage, markings, and illuminated exit signs indicating the nearest exit routes.

2. **Training:**

   Participate in safety orientation and emergency preparedness training sessions conducted by the cleanroom management team. These sessions typically include information on emergency exits, evacuation procedures, and response protocols.

3. **Practice Drills:**

   Regularly participate in emergency evacuation drills to reinforce knowledge of emergency exits and evacuation procedures. Practice identifying and using different exit routes under simulated emergency conditions.

4.  **Report Obstructions:**

    Immediately report any obstructions, blockages, or deficiencies observed at emergency exits to cleanroom management or facility maintenance personnel. Prompt resolution of such issues is essential for maintaining the effectiveness of emergency evacuation routes.

By recognizing the importance of emergency exits and taking proactive steps to familiarize themselves with their locations and use, cleanroom personnel can contribute to a safer and more secure work environment for themselves and their colleagues. Emergency preparedness is a shared responsibility that requires cooperation, awareness, and readiness to respond effectively to unforeseen emergencies.

## 2.1.4: Emergency Contacts

In addition to knowing the location of emergency exits and evacuation procedures, it is essential for cleanroom personnel to have access to a list of emergency contacts. This section highlights the importance of maintaining a readily available list of emergency contacts and provides guidance on whom to contact in various emergency scenarios.

**Importance of Emergency Contacts:**

1.  **Immediate Assistance:**

    During emergencies, quick access to emergency contacts can expedite the process of summoning assistance and initiating an appropriate response. Whether it's contacting emergency services, such as fire or medical assistance, or notifying cleanroom management and designated safety personnel, having a list of emergency contacts readily available is essential for prompt action.

2.  **Clarification of Procedures:**

    Emergency contacts can provide guidance and clarification on appropriate response procedures and protocols in different emergency situations. Cleanroom personnel may need to seek advice on evacuation routes, containment measures, or handling hazardous materials, depending on the nature of the emergency. Emergency contacts serve as valuable resources for obtaining accurate information and guidance under stress.

3.  **Coordination of Resources:**

    In complex or large-scale emergencies, coordination of resources and response efforts may be necessary to mitigate the situation effectively. Emergency contacts can facilitate communication and coordination among cleanroom personnel, emergency responders, facility management, and external agencies, ensuring a unified and coordinated response to the emergency.

4. **Notification of Next of Kin:**

In the event of a serious incident or injury involving cleanroom personnel, emergency contacts may include provisions for notifying next of kin or designated contacts. Timely notification of family members or guardians is crucial for providing support, reassurance, and assistance to affected individuals and their loved ones during challenging circumstances.

**Recommended Emergency Contacts:**

1. **Local Emergency Services:**

Include contact information for local emergency services, such as fire, police, and medical services. Dialing emergency services, such as 911 in the United States, should be the first step in response to life-threatening emergencies requiring immediate assistance.

2. **Cleanroom Management:**

Maintain contact information for cleanroom management, including supervisors, safety officers, and designated emergency response personnel. Cleanroom management personnel can provide guidance, assistance, and coordination of response efforts during cleanroom-related emergencies.

3. **Facility Security:**

Include contact details for facility security personnel or designated security officers responsible for managing access control, surveillance, and security-related matters within the cleanroom facility.

4. **Health and Safety Department:**

If applicable, include contact information for the health and safety department or occupational health services responsible for overseeing workplace safety, health assessments, and medical support for cleanroom personnel.

5. **Chemical Spill Response Team:**

For cleanrooms handling hazardous chemicals or materials, maintain contact information for the chemical spill response team or hazardous materials (HAZMAT) team responsible for managing chemical spills, containment, and cleanup procedures.

**Accessibility of Emergency Contacts:**

1. **Readily Available:**

   Ensure that the list of emergency contacts is readily available and easily accessible to all cleanroom personnel. Display the list prominently in designated areas within the cleanroom facility, such as near exits, entrances, or emergency equipment stations.

2. **Digital and Physical Copies:**

   Provide both digital and physical copies of the list of emergency contacts to cleanroom personnel. Digital copies can be stored on electronic devices, such as smartphones or tablets, while physical copies can be posted on notice boards or distributed as printed handouts.

3. **Regular Updates:**

   Review and update the list of emergency contacts regularly to ensure accuracy and relevance. Changes in personnel, contact information, or emergency response protocols should be promptly reflected in the updated list to maintain its effectiveness and reliability during emergencies.

By maintaining a list of emergency contacts and ensuring its accessibility to all cleanroom personnel, individuals can enhance preparedness, facilitate effective communication, and streamline response efforts during emergencies. Remember, swift action and clear communication are paramount in ensuring the safety and well-being of all individuals within the cleanroom environment.

## 2.2 Hazardous Material Handling:

## 2.2.1 Material Identification

Material identification in a cleanroom setting involves the process of recognizing and categorizing substances based on specific characteristics, labels, symbols, or documentation. This is a critical aspect of Hazardous Material Handling to ensure that personnel are aware of the nature of the materials they are dealing with. Here's an explanation of material identification in a cleanroom context:

1. **Labels and Symbols:**

   Containers or packages containing hazardous materials are labeled with standardized symbols and text to indicate their contents. Common symbols include those for flammable materials, corrosive substances, toxic compounds, etc. Labels provide immediate visual cues about the nature of the material.

2. **Material Safety Data Sheets (MSDS):**

   MSDS is a comprehensive document that provides detailed information about the properties, hazards, handling procedures, and emergency response measures for a specific material. Cleanroom personnel should be trained to read and understand MSDS to identify the properties and risks associated with each substance.

3. **Color Coding:**

   Some cleanroom facilities use color coding to differentiate between types of materials. For example, different colors may be assigned to containers based on the hazard level of the contents. This visual coding aids in quick identification.

4. **Inventory Management Systems:**

   Cleanrooms may employ inventory management systems that use barcodes or RFID technology to track and identify materials. These systems provide real-time information about the location and status of materials, enhancing visibility and control.

5. **Training and Education:**

   Cleanroom personnel undergo training to recognize and identify hazardous materials. This training includes familiarizing individuals with common labels, symbols, and procedures for identifying substances. Regular updates and refresher courses reinforce this knowledge.

6. **Supplier Documentation:**

   Materials received in a cleanroom often come with documentation from suppliers, including labels, safety information, and compliance certificates. Cleanroom personnel should rely on this documentation to understand the characteristics and hazards associated with the materials.

7. **Communication:**

   Effective communication is essential in a cleanroom environment. Personnel should be encouraged to communicate with each other regarding the identification of materials. This includes reporting any discrepancies in labeling or concerns about unidentified substances.

8. **Integration with Cleanroom Procedures:**

   Material identification is seamlessly integrated into cleanroom procedures. This includes incorporating identification checks into receiving processes, storage protocols, and transfer procedures. The identification process becomes a routine part of daily activities.

By employing a combination of these methods, cleanroom facilities ensure that hazardous materials are clearly identified, and personnel can take appropriate precautions in handling, storing, and disposing of these substances. Material identification is a foundational step in maintaining a safe and controlled cleanroom environment.

## 2.2.2 Handling Protocols

Handling protocols for hazardous materials in a cleanroom involve specific guidelines and procedures to ensure the safe and secure management of substances that pose potential risks to personnel, the environment, or the integrity of the cleanroom itself. Here's an explanation of handling protocols in a cleanroom setting:

1. **Access Control:**

   Limit access to designated personnel with proper training and authorization. Unauthorized individuals should not enter areas where hazardous materials are handled.

2. **Personal Protective Equipment (PPE):**

   Personnel must wear appropriate PPE as specified for the particular hazardous material being handled. This may include gloves, goggles, masks, lab coats, or specialized suits to prevent direct contact and inhalation.

3. **Training and Certification:**

   Cleanroom personnel involved in handling hazardous materials undergo thorough training programs. Certification may be required to ensure that individuals are well-versed in the specific protocols related to the materials they handle.

4. **Material Segregation:**

   Store and handle materials in accordance with compatibility guidelines. Segregate incompatible substances to prevent reactions and ensure the stability of the materials.

5. **Handling Tools and Equipment:**

Utilize specialized tools and equipment designed for the safe handling of hazardous materials. This may include dedicated containers, transfer systems, and equipment resistant to chemical reactions.

6. **Spill Response and Containment:**

Establish procedures for immediate response to spills or leaks. Personnel should be trained to contain and clean up spills following predefined protocols to minimize exposure and environmental impact.

7. **Ventilation and Containment Systems:**

Ensure that the cleanroom is equipped with adequate ventilation and containment systems to control the dispersion of hazardous fumes or particles. These systems help maintain air quality within acceptable limits.

8. **Emergency Procedures:**

Develop and communicate clear emergency procedures for the handling of hazardous materials. This includes evacuation plans, emergency shutdown protocols, and communication channels for reporting incidents.

9. **Material Transfer and Transport:**

Implement secure methods for transferring and transporting hazardous materials within the cleanroom. This may involve using sealed containers, carts, or conveyance systems that prevent accidental exposure or contamination.

10. **Waste Disposal:**

Dispose of hazardous waste according to regulatory requirements. Implement procedures for proper segregation, labeling, and disposal of materials to prevent environmental pollution and ensure compliance.

11. **Monitoring and Documentation:**

Regularly monitor handling activities and document all processes related to hazardous materials. This includes recording material transfers, usage, and any incidents. Documentation serves as a crucial reference for audits and continuous improvement.

**12. Regular Audits and Inspections:**

Conduct routine audits and inspections to assess compliance with handling protocols. Identify areas for improvement and address any deviations from established procedures.

By adhering to these handling protocols, cleanroom facilities can minimize the risks associated with hazardous materials, protect the well-being of personnel, and maintain the stringent cleanliness standards required in controlled environments.

## 2.2.3 Personal Protective Equipment (PPE):

Personal Protective Equipment (PPE) is essential for ensuring the safety of personnel when handling hazardous materials in a cleanroom setting. The specific PPE required can vary depending on the type of hazardous material being handled. Here are examples of suitable PPE for different types of materials:

- **Chemical Hazard Suitable PPE:**
    - Chemical-resistant gloves
    - Chemical splash goggles or face shield
    - Chemical-resistant apron or lab coat
    - Closed-toe, chemical-resistant shoes
- **Biological Hazard Suitable PPE:**
    - Disposable gloves
    - Face mask or respirator (if aerosols are a concern)
    - Protective clothing or disposable coveralls
    - Shoe covers
- **Radioactive Material Suitable PPE:**
    - Lead apron or full-body radiation protection suit
    - Lead gloves
    - Protective eyewear
    - Shoe covers
- **Particulate Hazard (Dust, Nanoparticles) Suitable PPE:**
    - N95 or higher-rated respirator mask

- Disposable coveralls or cleanroom suit

  - Shoe covers

  - Hairnet or full head covering

- **Sharp Objects Hazard Suitable PPE:**

  - Cut-resistant gloves

  - Long-sleeved shirts and pants

  - Steel-toed safety shoes

  - Safety glasses or goggles

- **Biological Contaminants (Microorganisms) Suitable PPE:**

  - Disposable gloves

  - Face mask or respirator

  - Full-body protective clothing or disposable coveralls

  - Shoe covers

- **Electrostatic Discharge (ESD) Hazard Suitable PPE:**

  - ESD-safe gloves

  - ESD-safe footwear (shoes with conductive soles)

  - ESD-safe coveralls or smock

  - Anti-static wrist strap or heel strap

- **High-Temperature Hazard Suitable PPE:**

  - Heat-resistant gloves

  - Heat-resistant face shield or goggles

  - Heat-resistant apron or coat

  - Insulated footwear

- **Cryogenic Hazard Suitable PPE:**

  - Cryogenic-resistant gloves

  - Face shield or safety goggles

  - Cryogenic-resistant apron or suit

- Insulated footwear

It is crucial to conduct a thorough risk assessment to determine the specific hazards associated with the materials being handled in the cleanroom. The selection of appropriate PPE should align with the identified risks, and personnel should be trained on the correct usage, limitations, and maintenance of their PPE. Regular reviews of PPE protocols and updates based on changes in materials or processes are also essential to ensure ongoing safety and compliance.

## 2.2.4 Spill Response:

Spill response in a cleanroom involves a set of procedures and actions to quickly and effectively address accidental spills of hazardous materials. The goal is to minimize the impact of the spill on personnel safety, the environment, and the integrity of the cleanroom. Here's an explanation of spill response in a cleanroom setting:

1. **Immediate Action:**

   Upon discovering a spill, personnel must take immediate action to prevent the spread of the hazardous material. This may involve activating emergency alarms, notifying others in the vicinity, and securing the area.

2. **Personal Protective Equipment (PPE):**

   Responders should wear appropriate PPE to protect themselves from exposure to the spilled material. Examples of PPE for spill response may include:

   - Chemical-resistant gloves

   - Goggles or face shields

   - Respirators, if airborne contaminants are present

   - Protective suits or aprons

3. **Evacuation and Isolation:**

   In some cases, it may be necessary to evacuate personnel from the affected area to a safe location. Isolating the spill area helps prevent further contamination and exposure.

4. **Containment:**

   Use appropriate containment measures to prevent the spread of the spilled material. This may include using absorbent materials like spill kits, barriers, or absorbent socks to confine the spill.

5. **Spill Kit Utilization:**

   Cleanroom facilities should be equipped with spill kits tailored to the types of materials used. These kits typically include absorbents, neutralizers, containment tools, and PPE necessary for responding to spills.

6. **Material-Specific Response:**

   Depending on the nature of the spilled material, responders should follow material-specific response procedures. For example:

   - **Chemical Spills:**

     Use appropriate neutralizing agents or absorbents to safely clean up the spill.

   - **Biological Spills:**

     Follow containment and disinfection protocols to minimize the risk of contamination.

   - **Radioactive Spills:**

     Implement procedures for handling radioactive materials, including shielding and proper disposal.

7. **Ventilation and Airflow Management:**

   Adjust ventilation systems to control the dispersion of airborne contaminants. This helps maintain cleanroom air quality during and after the spill response.

8. **Waste Disposal:**

   Dispose of contaminated materials and absorbents according to regulatory guidelines. Clearly label and segregate waste for proper disposal.

9. **Decontamination:**

   After the spill is contained and cleaned up, decontaminate affected surfaces, tools, and equipment. Follow established protocols to ensure that the cleanroom environment is restored to its required level of cleanliness.

**10.Documentation:**

> Record all details of the spill response, including the type and quantity of the spilled material, actions taken, and personnel involved. This documentation is crucial for internal reviews, regulatory compliance, and continuous improvement.

Spill response procedures should be well-documented, regularly reviewed, and practiced through drills to ensure that cleanroom personnel are prepared to handle such incidents swiftly and safely.

## 2.3 First Aid Protocols

## 2.3.1 First Aid Station

The first aid station in a cleanroom is a designated area equipped to provide immediate and basic medical assistance to individuals who may sustain injuries or experience health issues while working in the cleanroom environment. Here's an explanation of the components and functions of a first aid station in a cleanroom:

1. **Location:**

   > The first aid station is strategically located within or near the cleanroom facility, ensuring quick and easy access for cleanroom personnel. The location is typically marked and known to all occupants.

2. **Components:**

   > The first aid station is stocked with essential first aid supplies and equipment. Common components include:
   >
   > - Sterile bandages and dressings
   > - Antiseptic wipes or solutions
   > - Adhesive tape
   > - Scissors and tweezers
   > - Disposable gloves
   > - Instant cold packs
   > - Pain relievers (as allowed by cleanroom policies)
   > - Eye wash station or solution
   > - CPR face shield
   > - Emergency contact information

3. **First Aid Personnel:**

   Trained individuals or designated first aid responders are assigned to the cleanroom first aid station. These personnel have basic first aid training and are familiar with cleanroom protocols.

4. **Communication:**

   - **Explanation:** The first aid station is equipped with communication tools to quickly alert emergency response teams, if necessary. This may include emergency phones, intercom systems, or other communication devices.

5. **Emergency Response Plan:**

   - **Explanation:** The first aid station is part of the overall cleanroom emergency response plan. It outlines the procedures to be followed in the event of injuries or medical emergencies, including how to summon additional medical help if required.

6. **Documentation:**

   All first aid incidents are documented, including the nature of the injury or illness, the treatment provided, and any follow-up recommendations. This documentation is crucial for internal reporting, regulatory compliance, and continuous improvement.

7. **Training and Drills:**

   Personnel assigned to the first aid station undergo regular training and participate in emergency response drills. This ensures that they are well-prepared to handle various medical situations and that their responses align with cleanroom safety protocols.

8. **Maintaining Cleanroom Standards:**

   While providing medical assistance, the first aid personnel take measures to maintain cleanroom cleanliness. This includes using appropriate protective gear, ensuring that any waste generated is properly disposed of, and minimizing the risk of contamination.

9. **Accessibility:**

   The first aid station is easily accessible, and its location is known to all cleanroom personnel. Clear signage directs individuals to the first aid station, contributing to a rapid response in case of emergencies.

10. **Continuous Improvement:**

> The first aid station's procedures and supplies are regularly reviewed and updated based on incident reports, regulatory changes, and advancements in first aid practices. This continuous improvement ensures that the first aid station remains effective in addressing medical needs within the cleanroom.

Overall, the first aid station plays a crucial role in promoting the health and safety of cleanroom personnel, providing timely and appropriate medical care when needed.

## 2.3.2 Basic First Aid Skills

Basic first aid skills are fundamental abilities and knowledge that individuals should possess to provide immediate and initial care in the event of an injury or sudden illness. These skills are essential in various settings, including cleanrooms, to address medical emergencies until professional medical help arrives. Here's an explanation of basic first aid skills:

1. **Assessment of the Situation:**

   > The first aider assesses the situation to ensure personal safety and identify potential hazards. This includes evaluating the need for additional assistance and assessing the nature and severity of injuries or illnesses.

2. **Primary Survey (ABCs):**

   > The primary survey involves checking the Airway, Breathing, and Circulation (ABCs) to ensure that the injured or ill person is breathing, has an open airway, and has a circulating heartbeat. Immediate interventions are applied if any of these are compromised.

3. **Calling for Help:**

   > If the situation requires professional medical assistance, the first aider calls for help, alerting emergency services or designated personnel in accordance with established protocols.

4. **Cardiopulmonary Resuscitation (CPR):**

   > Basic knowledge of CPR is crucial. This includes chest compressions and rescue breaths for individuals who are unresponsive and not breathing. AED (Automated External Defibrillator) use may also be included in CPR training.

5. **Control of Bleeding:**

   The ability to control bleeding using direct pressure, elevation, and the application of bandages or dressings. It also includes recognizing severe bleeding that may require immediate attention.

6. **Treatment of Shock:**

   Understanding and applying measures to treat shock, including maintaining a person's body temperature, keeping them calm, and ensuring proper circulation.

7. **Wound Care:**

   Cleaning and dressing minor wounds to prevent infection. This includes knowledge of basic wound care principles, such as cleaning with mild soap and water and applying sterile dressings.

8. **Burn Management:**

   Providing first aid for burns, including cooling the affected area with cold water, covering the burn with a clean cloth, and avoiding the use of ice.

9. **Choking Response:**

   Knowing how to respond to a choking person, including performing abdominal thrusts (Heimlich maneuver) to clear the airway.

10. **Basic Bandaging:**

    Applying basic bandages and dressings to wounds and injuries. This includes knowledge of different types of bandages and their appropriate uses.

11. **Fracture and Sprain Management:**

    Recognizing and providing basic first aid for fractures and sprains. This may involve immobilizing the injured area and seeking professional medical help.

12. **Allergic Reaction Management:**

    Recognizing the signs of allergic reactions and administering basic care, such as administering an epinephrine auto-injector for severe allergic reactions (anaphylaxis).

13. **Basic Medication Administration:**

Administering basic over-the-counter medications as appropriate and as allowed by local regulations and policies.

14. **Communication and Reassurance:**

Effectively communicating with the injured or ill person, providing reassurance, and maintaining a calm and supportive demeanor.

15. **Documentation:**

Documenting the first aid provided, including details of the incident, actions taken, and any recommendations for follow-up care.

Basic first aid skills empower individuals to respond effectively in emergency situations, potentially preventing further harm and improving outcomes until professional medical assistance is available. Training and regular refreshers in these skills are essential for cleanroom personnel to ensure a safe and prepared work environment.

## 2.3.3 Emergency Response Teams:

Emergency Response Teams (ERTs) are groups of trained individuals within an organization or facility responsible for responding to emergencies, providing immediate assistance, and managing critical situations. The purpose of ERTs is to enhance the overall safety and preparedness of the workplace. Here's an explanation of Emergency Response Teams and their key components:

1. **Formation and Composition:**

   Emergency Response Teams are formed by selecting individuals from various departments or units within an organization. Team members are trained to handle specific types of emergencies, ensuring a diverse skill set within the team.

2. **Training and Certification:**

   ERT members undergo specialized training in emergency response procedures, first aid, CPR, the use of firefighting equipment, evacuation protocols, and other relevant skills. Certification is often required to ensure competency.

3. **Roles and Responsibilities:**

   Each member of the Emergency Response Team has specific roles and responsibilities based on their training. This may include first aid response, fire suppression, evacuation coordination, communication management, and other tasks as defined by the organization's emergency response plan.

4. **Emergency Response Plan Implementation:**

   ERTs play a crucial role in implementing the organization's emergency response plan. They follow established protocols and procedures to address various emergencies, such as fires, medical incidents, chemical spills, natural disasters, or other critical events.

5. **First Aid and Medical Assistance:**

   ERT members are trained in first aid and medical response. They can provide immediate assistance to individuals who are injured or unwell until professional medical help arrives. This includes basic life support and the use of automated external defibrillators (AEDs).

6. **Firefighting and Evacuation:**

   ERTs are equipped to handle small fires using firefighting equipment like extinguishers. They also play a key role in coordinating and assisting with evacuation procedures during fire alarms or other emergencies requiring the evacuation of personnel.

7. **Communication and Coordination:**

   Effective communication is critical during emergencies. ERTs are trained to use communication tools and systems to relay information to all personnel, coordinate responses, and liaise with external emergency services.

8. **Drills and Exercises:**

   Regular drills and exercises are conducted to ensure that ERT members are familiar with their roles and can respond efficiently to various emergency scenarios. These drills may include fire drills, medical emergency simulations, and other relevant exercises.

9. **Equipment Management:**

   ERTs are responsible for the maintenance and accessibility of emergency response equipment. This includes first aid kits, firefighting equipment, communication devices, and any other tools required for emergency response.

10. **Continuous Improvement:**

> ERTs participate in ongoing training and continuous improvement initiatives. They review the outcomes of drills, assess their responses to actual emergencies, and update their skills and procedures as needed.

11. **Coordination with External Services:**

> In certain situations, ERTs may need to coordinate with external emergency services such as fire departments, paramedics, or hazardous material response teams. Effective communication and collaboration are essential.

Emergency Response Teams are a vital component of workplace safety, contributing to a swift and organized response to emergencies. Their presence and readiness help mitigate risks and protect the well-being of individuals within the organization.

## 2.4 Personal Hygiene Practices:

## 2.4.1 Handwashing

Handwashing is a fundamental and effective practice in maintaining personal hygiene and preventing the spread of infections, especially in cleanroom environments. Here is an explanation of handwashing in the context of personal hygiene practices:

1. **Importance of Handwashing:**

   - **Explanation:** Handwashing is crucial for preventing the transfer of bacteria, viruses, and other contaminants from hands to surfaces, equipment, or other individuals. In cleanroom environments, where the control of particulate and microbial contamination is paramount, proper hand hygiene is essential.

2. **When to Wash Hands:**

   - **Explanation:** Hands should be washed at specific times, including before entering the cleanroom, after using the restroom, before handling sensitive equipment, after touching the face or body, and after handling any potentially contaminated materials. Adhering to a strict handwashing schedule is vital.

3. **Proper Handwashing Technique:**

   - **Explanation:** The technique for handwashing is critical. Individuals should wet their hands with clean, running water, apply soap, and scrub all surfaces of the hands, including the back of the hands, between the fingers, and under the nails, for at least 20 seconds. Thorough rinsing with water and proper drying with disposable towels are also important.

4. **Use of Approved Soaps and Disinfectants:**

   - **Explanation:** In cleanroom environments, it is essential to use soaps and hand sanitizers approved for use in controlled environments. These products should be selected to minimize the risk of introducing contaminants and to maintain the integrity of the cleanroom environment.

5. **Nail and Hand Jewelry Maintenance:**

   - **Explanation:** Individuals working in cleanrooms should keep their nails short and clean to prevent the accumulation of dirt and contaminants. Additionally, the wearing of hand jewelry, such as rings and watches, may be restricted to reduce the risk of harboring microbes.

6. **Hand Drying:**

   - **Explanation:** After washing, hands should be dried thoroughly using disposable, lint-free towels or air dryers. Damp hands can attract and transfer contaminants more easily than dry hands.

7. **Avoiding Touching Contaminated Surfaces:**

   - **Explanation:** Individuals should be mindful of touching surfaces that may be contaminated. If contact occurs, handwashing should be performed immediately. This is particularly important in cleanroom environments where even slight contamination can be significant.

8. **Education and Training:**

   - **Explanation:** Proper handwashing techniques should be part of the education and training provided to personnel working in cleanrooms. Regular reminders and reinforcement of these practices contribute to a culture of strict adherence to hygiene protocols.

9. **Regular Monitoring and Audits:**

   - **Explanation:** Cleanroom facilities may conduct regular monitoring and audits to ensure compliance with hand hygiene practices. This may involve visual inspections, microbial testing, and other measures to verify the effectiveness of handwashing procedures.

10. **Documentation and Reporting:**

- **Explanation:** Cleanroom protocols often include documentation of handwashing activities. Individuals may be required to record their handwashing activities, and this information can be valuable for tracking and addressing any deviations from hygiene protocols.

Maintaining impeccable hand hygiene is a shared responsibility among all individuals working in cleanroom environments. By following rigorous handwashing practices, personnel contribute to the overall cleanliness and integrity of the controlled environment, minimizing the risk of contamination.

## 2.4.2 Use of Personal Care Products:

Maintaining cleanliness and preventing contamination in cleanroom environments extend beyond work attire to include personal care products. Here's an explanation of the guidelines on using non-contaminating personal care products for cleanroom users:

1. **Product Selection:**

   - **Explanation:** Cleanroom users should select personal care products that are specifically designed and labeled as suitable for cleanroom environments. These products are formulated to minimize the introduction of contaminants, such as particles and residues, that could compromise the controlled environment.

2. **Fragrance-Free and Dye-Free Formulas:**

   - **Explanation:** Personal care products, including lotions, soaps, and shampoos, should preferably be fragrance-free and dye-free. Fragrances and dyes may contain volatile organic compounds (VOCs) that can contribute to airborne contamination within the cleanroom.

3. **Low-Residue Formulations:**

   - **Explanation:** Products with low-residue formulations are preferred to reduce the risk of leaving behind particles on surfaces or equipment. Residue from personal care products could potentially contaminate critical workspaces and affect the integrity of processes.

4. **Alcohol-Based Hand Sanitizers:**

   - **Explanation:** When hand sanitizers are used, alcohol-based formulations are often recommended. These sanitizers are effective at reducing microbial contamination on hands. However, the alcohol content should be compatible with cleanroom standards and not introduce additional contaminants.

5. **Approved Disinfectants and Wipes:**

   - **Explanation:** Disinfectants and wipes used for personal hygiene or surface cleaning should be approved for cleanroom use. These products should be selected based on their ability to eliminate contaminants without compromising the controlled environment.

6. **Avoidance of Excessive Moisturizers:**

   - **Explanation:** Moisturizers, if used, should be applied sparingly to avoid excessive residue. Some moisturizers may contain ingredients that can contribute to particle generation or create a film on surfaces, impacting the cleanliness of the cleanroom.

7. **Hair Care Products:**

   - **Explanation:** Hair care products such as shampoos and conditioners should meet cleanroom standards. These products should be chosen for their low-particulate and low-residue characteristics to prevent contamination of workspaces and equipment.

8. **Guidelines for Makeup Wearers:**

   - **Explanation:** Cleanroom users who wear makeup should choose products with minimal shedding or flaking properties. Loose particles from makeup can become airborne and settle on critical surfaces, posing a risk to the cleanroom environment.

9. **Regular Training and Awareness:**

   - **Explanation:** Cleanroom users should receive regular training and awareness sessions regarding the selection and proper use of personal care products. This education ensures that individuals are informed about the potential impact of these products on cleanroom operations.

10. **Periodic Reviews of Product Suitability:**

    - **Explanation:** The suitability of personal care products should be periodically reviewed in light of any changes in formulations, product availability, or cleanroom requirements. This ensures ongoing compliance with cleanliness standards.

11. **Documentation and Reporting:**

- **Explanation:** Cleanroom protocols may include documentation of personal care product usage. Individuals may be required to report any adverse effects or deviations from guidelines, contributing to continuous improvement in personal care practices.

By adhering to guidelines for the use of non-contaminating personal care products, cleanroom users contribute to the maintenance of a controlled and sterile environment, supporting the precision and integrity of processes conducted within cleanroom facilities.

## 2.4.3 Clothing and Jewelry:

Maintaining a high level of cleanliness in a controlled environment like a cleanroom involves specific guidelines for clothing and jewelry. Here's an explanation of the key considerations to ensure proper attire and minimal use of accessories in a cleanroom:

1. **Cleanroom Garments:**

   - **Explanation:** Individuals entering a cleanroom should wear dedicated cleanroom garments, including coveralls, hoods, and booties. These garments are designed to minimize the shedding of particles and contaminants. The use of cleanroom-specific attire helps maintain the cleanliness of the controlled environment.

2. **Material Selection:**

   - **Explanation:** Cleanroom garments should be made from materials that are low-linting and do not generate excessive particles. Common materials include synthetic fabrics like polyester or other approved materials that meet cleanroom standards.

3. **No Exposed Skin:**

   - **Explanation:** To prevent the shedding of skin particles, cleanroom attire should cover the entire body. Exposed skin can introduce contaminants into the cleanroom environment. Full coverage minimizes the risk of particles originating from the body.

4. **Head Coverings:**

   - **Explanation:** Head coverings, such as hoods or bouffant caps, are essential to prevent hair and skin particles from entering the cleanroom environment. Cleanroom personnel should ensure that head coverings are worn consistently and securely.

5. **Minimize Accessories:**

   - **Explanation:** Individuals entering a cleanroom should minimize the use of accessories such as jewelry, watches, and other personal items. These accessories can contribute to the generation of particles or harbor contaminants that may compromise the cleanliness of the controlled environment.

6. **Approved Cleanroom Shoes:**

   - **Explanation:** Cleanroom shoes or shoe covers should be worn to prevent the introduction of contaminants from footwear. The shoes should be made from materials that do not shed particles and are easy to clean. Shoe covers provide an additional layer of protection.

7. **Prohibited Items:**

   - **Explanation:** Certain items, such as pens, notebooks, and non-essential tools, may introduce contaminants. Cleanroom protocols often prohibit the entry of such items, and individuals should be aware of and adhere to these restrictions.

8. **Jewelry Restrictions:**

   - **Explanation:** Cleanroom guidelines typically restrict the wearing of jewelry. Rings, bracelets, necklaces, and other accessories can harbor particles, pose a cleanliness risk, and interfere with the proper wearing of cleanroom garments.

9. **Gloves Usage:**

   - **Explanation:** Depending on the cleanroom classification and specific processes, gloves may be required. Proper glove usage helps prevent the transfer of particles from hands to critical surfaces or equipment.

10. **Regular Change of Garments:**

    - **Explanation:** Cleanroom personnel should change into cleanroom garments upon entering the controlled environment. Regular changes, especially after breaks or exits from the cleanroom, help maintain the desired level of cleanliness.

11. **Training and Compliance:**

    - **Explanation:** Individuals working in cleanrooms should receive training on proper attire and adherence to cleanroom protocols. Regular compliance checks and reminders contribute to a culture of cleanliness within the cleanroom facility.

By following these clothing and jewelry guidelines, cleanroom users contribute to the overall cleanliness and integrity of the controlled environment, ensuring that critical processes are conducted without the risk of contamination.

## 2.5 Cleanroom Behavior:

### 2.5.1 Restricted Behaviors:

Maintaining a cleanroom's controlled environment requires strict adherence to specific behaviors to prevent contamination. Here's an explanation of restricted behaviors, outlining activities that are not allowed within a cleanroom setting:

1. **Eating:**

   - **Explanation:** Consuming food in a cleanroom is strictly prohibited. Food particles, crumbs, or spills can introduce contaminants into the environment, compromising the cleanliness required for sensitive processes and equipment.

2. **Drinking:**

   - **Explanation:** Drinking, including beverages in open containers, is not allowed in a cleanroom. Liquids can easily spill, and open containers pose a risk of introducing contaminants. Cleanroom personnel should hydrate outside the cleanroom environment.

3. **Smoking or Vaping:**

   - **Explanation:** Smoking or vaping is strictly prohibited in cleanrooms. The release of smoke or vapor introduces particulate matter and chemicals that can compromise the cleanroom's integrity.

4. **Chewing Gum:**

   - **Explanation:** Chewing gum is not allowed in a cleanroom. Gum particles can be released into the air, and improper disposal poses a contamination risk. Cleanroom users should refrain from chewing gum while inside the controlled environment.

5. **Talking Closely or Sneezing/Coughing:**

   - **Explanation:** Maintaining a safe distance while conversing is important to prevent the transfer of respiratory particles. Sneezing or coughing without proper containment measures, such as using tissues or following hygiene protocols, is restricted to avoid the spread of contaminants.

6. **Touching Surfaces Unnecessarily:**

   - **Explanation:** Cleanroom users should avoid unnecessary contact with surfaces, equipment, or tools. Touching surfaces can transfer oils, skin particles, or contaminants, impacting the cleanliness of critical areas.

7. **Using Personal Items:**

   - **Explanation:** Bringing personal items, such as non-essential tools, pens, or personal belongings, into the cleanroom is restricted. These items can introduce contaminants and should be stored outside the cleanroom or in designated areas.

8. **Wearing Personal Items Improperly:**

   - **Explanation:** Cleanroom personnel should wear cleanroom attire properly, ensuring that hoods, masks, and other garments are correctly fitted. Wearing attire improperly can compromise its effectiveness in preventing particle shedding.

9. **Bringing Unauthorized Equipment:**

   - **Explanation:** Only equipment approved for cleanroom use should be brought inside. Unauthorized tools or equipment may introduce contaminants or disrupt the controlled environment. Cleanroom users should be familiar with the approved list of tools.

10. **Engaging in Unsafe Practices:**

    - **Explanation:** Any behavior that poses a safety risk, including actions that could result in spills, accidents, or contamination, is strictly restricted. Cleanroom users should adhere to safety protocols and report any unsafe conditions.

11. **Ignoring Cleanroom Entry Procedures:**

    - **Explanation:** Proper entry procedures, including gowning and hygiene practices, must be followed. Ignoring or bypassing these procedures can compromise the cleanliness of the cleanroom.

12. **Failing to Report Contamination Incidents:**

    - **Explanation:** Cleanroom users must promptly report any incidents of contamination, spills, or breaches of cleanroom protocols. Timely reporting allows for quick remediation and minimizes the impact on cleanroom processes.

Enforcing and communicating these restricted behaviors is essential for sustaining the desired cleanliness levels in a cleanroom environment. Strict adherence to these guidelines helps ensure the success of critical processes and the integrity of sensitive equipment within the controlled setting.

## 2.5.2 Prohibited Items:

Maintaining the cleanliness and controlled environment of a cleanroom requires strict regulations regarding items that are prohibited from being brought inside. Here is an explanation of prohibited items, emphasizing the items that should not be introduced into a cleanroom setting:

1. **Street Clothes:**

    - **Explanation:** Cleanrooms have specific gowning requirements to minimize particle shedding. Street clothes, which may carry lint, fibers, or contaminants from the external environment, are strictly prohibited.

2. **Non-Cleanroom Approved Shoes:**

    - **Explanation:** Shoes can track in dirt, debris, and contaminants from outside. Only cleanroom-approved footwear, often with smooth soles and no external particles, is allowed to maintain a controlled environment.

3. **Non-Essential Personal Items:**

    - **Explanation:** Bringing unnecessary personal items, such as bags, backpacks, or accessories, into the cleanroom is prohibited. These items may carry contaminants and pose a risk to the controlled environment.

4. **Outside Tools or Equipment:**

    - **Explanation:** Tools or equipment that have not been approved for cleanroom use are strictly prohibited. Unauthorized tools can introduce particles, oils, or substances that may compromise sensitive processes.

5. **Unapproved Writing Instruments:**

    - **Explanation:** Only cleanroom-approved writing instruments should be used inside the cleanroom. Pens or markers that generate particles or contaminants are not allowed.

6. **Makeup and Cosmetics:**

   - **Explanation:** Makeup and cosmetics can release particles into the air, potentially compromising the cleanroom environment. Cleanroom users should refrain from wearing makeup or use only approved, low-particle-emitting products.

7. **Food and Beverages:**

   - **Explanation:** Bringing in any type of food or beverages is strictly prohibited. Even sealed items can pose a risk of spills or contamination. Cleanroom users should consume food and beverages only in designated areas outside the cleanroom.

8. **Jewelry and Accessories:**

   - **Explanation:** Excessive jewelry, watches, or accessories are prohibited as they may shed particles. Cleanroom attire should be minimal, and any accessories worn should be cleanroom-approved.

9. **Loose Hair:**

   - **Explanation:** Loose hair can contribute to particle generation. Cleanroom users should secure their hair using approved methods, such as hairnets, hoods, or caps, to prevent contamination.

10. **Electronic Devices:**

    - **Explanation:** While some cleanrooms may allow specific electronic devices, others may prohibit them entirely. Devices should be cleanroom-approved and free from contaminants that could be released into the environment.

11. **Fragrances and Perfumes:**

    - **Explanation:** Strong fragrances and perfumes can release particles or chemicals into the air. Cleanroom users should refrain from using strongly scented products to avoid contamination.

12. **Unapproved Documentation:**

    - **Explanation:** Any documents or paperwork brought into the cleanroom should be cleanroom-approved to prevent the introduction of paper particles or ink contaminants.

13. **Glass or Ceramics:**

> Glass or ceramic items can break, leading to contamination risks. Only cleanroom-approved materials should be used within the controlled environment.

Strict enforcement of these prohibitions is crucial for maintaining the cleanliness and integrity of a cleanroom. By adhering to these guidelines, cleanroom users contribute to the success of critical processes and ensure the reliability of sensitive equipment within the controlled environment.

## 2.5.3 Reporting Health Issues:

Ensuring the health and well-being of personnel working in a cleanroom is paramount to maintaining a controlled and contamination-free environment. Establishing clear protocols for reporting health issues is essential. Here's an explanation of the importance and guidelines for reporting health issues in a cleanroom:

1. **Prompt Identification of Health Issues:**

   Cleanroom personnel must promptly identify and report any health issues they may be experiencing. Early identification allows for timely intervention and prevents the potential spread of illnesses.

2. **Individual Responsibility:**

   Each individual working in the cleanroom bears the responsibility of monitoring their health. This includes being vigilant about symptoms such as coughing, sneezing, skin rashes, or any signs of illness.

3. **Immediate Reporting to Supervisors:**

   Cleanroom users should report any health issues immediately to their supervisors or designated personnel responsible for health and safety. Timely reporting enables the implementation of appropriate measures to prevent the spread of illness within the cleanroom.

4. **Confidentiality and Privacy:**

   Encourage a culture of confidentiality and privacy regarding health issues. Individuals should feel comfortable reporting health concerns without fear of judgment. This promotes open communication and ensures that necessary actions are taken without compromising the affected individual's privacy.

5.  **Isolation and Temporary Removal:**

If an individual reports symptoms of illness, protocols should be in place for isolating them from the cleanroom environment. Temporary removal from the cleanroom may be necessary to prevent potential contamination of sensitive processes or equipment.

6.  **Medical Assessment:**

Individuals reporting health issues may be required to undergo a medical assessment. This can help determine the nature of the illness and assess whether the individual is fit to return to work in the cleanroom.

7.  **Return-to-Work Clearance:**

Clear guidelines should be established for when individuals can return to work in the cleanroom after experiencing a health issue. Medical clearance may be required to ensure that the individual is no longer a risk to the controlled environment.

8.  **Education and Training:**

Conduct regular education and training sessions on the importance of reporting health issues. This includes providing information on common symptoms, the significance of early reporting, and the collective responsibility of all cleanroom users in maintaining a healthy environment.

9.  **Hygiene Practices:**

Reinforce hygiene practices as a preventive measure. This includes proper handwashing, the use of personal protective equipment (PPE), and adherence to cleanroom gowning procedures to minimize the risk of spreading illnesses.

10. **Monitoring and Documentation:**

Establish a system for monitoring reported health issues and documenting the actions taken. This documentation contributes to ongoing improvements in health and safety protocols within the cleanroom.

By fostering a culture of prompt reporting and addressing health issues, cleanroom facilities can safeguard both personnel and the integrity of critical processes. These measures contribute to the overall success of cleanroom operations and the reliability of sensitive equipment within the controlled environment.

## 2.5.4 Restricted Behaviors:

Maintaining a cleanroom's controlled environment requires strict adherence to specific behaviors to prevent contamination. Here's an explanation of restricted behaviors, outlining activities that are not allowed within a cleanroom setting:

13. **Eating:**

    Consuming food in a cleanroom is strictly prohibited. Food particles, crumbs, or spills can introduce contaminants into the environment, compromising the cleanliness required for sensitive processes and equipment.

14. **Drinking:**

    Drinking, including beverages in open containers, is not allowed in a cleanroom. Liquids can easily spill, and open containers pose a risk of introducing contaminants. Cleanroom personnel should hydrate outside the cleanroom environment.

15. **Smoking or Vaping:**

    Smoking or vaping is strictly prohibited in cleanrooms. The release of smoke or vapor introduces particulate matter and chemicals that can compromise the cleanroom's integrity.

16. **Chewing Gum:**

    Chewing gum is not allowed in a cleanroom. Gum particles can be released into the air, and improper disposal poses a contamination risk. Cleanroom users should refrain from chewing gum while inside the controlled environment.

17. **Talking Closely or Sneezing/Coughing:**

    Maintaining a safe distance while conversing is important to prevent the transfer of respiratory particles. Sneezing or coughing without proper containment measures, such as using tissues or following hygiene protocols, is restricted to avoid the spread of contaminants.

18. **Touching Surfaces Unnecessarily:**

Cleanroom users should avoid unnecessary contact with surfaces, equipment, or tools. Touching surfaces can transfer oils, skin particles, or contaminants, impacting the cleanliness of critical areas.

19. **Using Personal Items:**

Bringing personal items, such as non-essential tools, pens, or personal belongings, into the cleanroom is restricted. These items can introduce contaminants and should be stored outside the cleanroom or in designated areas.

20. **Wearing Personal Items Improperly:**

Cleanroom personnel should wear cleanroom attire properly, ensuring that hoods, masks, and other garments are correctly fitted. Wearing attire improperly can compromise its effectiveness in preventing particle shedding.

21. **Bringing Unauthorized Equipment:**

Only equipment approved for cleanroom use should be brought inside. Unauthorized tools or equipment may introduce contaminants or disrupt the controlled environment. Cleanroom users should be familiar with the approved list of tools.

22. **Engaging in Unsafe Practices:**

Any behavior that poses a safety risk, including actions that could result in spills, accidents, or contamination, is strictly restricted. Cleanroom users should adhere to safety protocols and report any unsafe conditions.

23. **Ignoring Cleanroom Entry Procedures:**

Proper entry procedures, including gowning and hygiene practices, must be followed. Ignoring or bypassing these procedures can compromise the cleanliness of the cleanroom.

24. **Failing to Report Contamination Incidents:**

Cleanroom users must promptly report any incidents of contamination, spills, or breaches of cleanroom protocols. Timely reporting allows for quick remediation and minimizes the impact on cleanroom processes.

Enforcing and communicating these restricted behaviors is essential for sustaining the desired cleanliness levels in a cleanroom environment. Strict adherence to these guidelines helps ensure the success of critical processes and the integrity of sensitive equipment within the controlled setting.

## 2.5.5 Prohibited Items:

Maintaining the cleanliness and controlled environment of a cleanroom requires strict regulations regarding items that are prohibited from being brought inside. Here is an explanation of prohibited items, emphasizing the items that should not be introduced into a cleanroom setting:

14. **Street Clothes:**

     Cleanrooms have specific gowning requirements to minimize particle shedding. Street clothes, which may carry lint, fibers, or contaminants from the external environment, are strictly prohibited.

15. **Non-Cleanroom Approved Shoes:**

     Shoes can track in dirt, debris, and contaminants from outside. Only cleanroom-approved footwear, often with smooth soles and no external particles, is allowed to maintain a controlled environment.

16. **Non-Essential Personal Items:**

     Bringing unnecessary personal items, such as bags, backpacks, or accessories, into the cleanroom is prohibited. These items may carry contaminants and pose a risk to the controlled environment.

17. **Outside Tools or Equipment:**

     Tools or equipment that have not been approved for cleanroom use are strictly prohibited. Unauthorized tools can introduce particles, oils, or substances that may compromise sensitive processes.

18. **Unapproved Writing Instruments:**

     Only cleanroom-approved writing instruments should be used inside the cleanroom. Pens or markers that generate particles or contaminants are not allowed.

19. **Makeup and Cosmetics:**

Makeup and cosmetics can release particles into the air, potentially compromising the cleanroom environment. Cleanroom users should refrain from wearing makeup or use only approved, low-particle-emitting products.

20. **Food and Beverages:**

Bringing in any type of food or beverages is strictly prohibited. Even sealed items can pose a risk of spills or contamination. Cleanroom users should consume food and beverages only in designated areas outside the cleanroom.

21. **Jewelry and Accessories:**

Excessive jewelry, watches, or accessories are prohibited as they may shed particles. Cleanroom attire should be minimal, and any accessories worn should be cleanroom-approved.

22. **Loose Hair:**

Loose hair can contribute to particle generation. Cleanroom users should secure their hair using approved methods, such as hairnets, hoods, or caps, to prevent contamination.

23. **Electronic Devices:**

While some cleanrooms may allow specific electronic devices, others may prohibit them entirely. Devices should be cleanroom-approved and free from contaminants that could be released into the environment.

24. **Fragrances and Perfumes:**

Strong fragrances and perfumes can release particles or chemicals into the air. Cleanroom users should refrain from using strongly scented products to avoid contamination.

25. **Unapproved Documentation:**

Any documents or paperwork brought into the cleanroom should be cleanroom-approved to prevent the introduction of paper particles or ink contaminants.

26. **Glass or Ceramics:**

> Glass or ceramic items can break, leading to contamination risks. Only cleanroom-approved materials should be used within the controlled environment.

Strict enforcement of these prohibitions is crucial for maintaining the cleanliness and integrity of a cleanroom. By adhering to these guidelines, cleanroom users contribute to the success of critical processes and ensure the reliability of sensitive equipment within the controlled environment.

## 2.5.6 Reporting Health Issues:

Ensuring the health and well-being of personnel working in a cleanroom is paramount to maintaining a controlled and contamination-free environment. Establishing clear protocols for reporting health issues is essential. Here's an explanation of the importance and guidelines for reporting health issues in a cleanroom:

11. **Prompt Identification of Health Issues:**

> Cleanroom personnel must promptly identify and report any health issues they may be experiencing. Early identification allows for timely intervention and prevents the potential spread of illnesses.

12. **Individual Responsibility:**

> Each individual working in the cleanroom bears the responsibility of monitoring their health. This includes being vigilant about symptoms such as coughing, sneezing, skin rashes, or any signs of illness.

13. **Immediate Reporting to Supervisors:**

> Cleanroom users should report any health issues immediately to their supervisors or designated personnel responsible for health and safety. Timely reporting enables the implementation of appropriate measures to prevent the spread of illness within the cleanroom.

14. **Confidentiality and Privacy:**

> Encourage a culture of confidentiality and privacy regarding health issues. Individuals should feel comfortable reporting health concerns without fear of judgment. This promotes open communication and ensures that necessary actions are taken without compromising the affected individual's privacy.

15. **Isolation and Temporary Removal:**

If an individual reports symptoms of illness, protocols should be in place for isolating them from the cleanroom environment. Temporary removal from the cleanroom may be necessary to prevent potential contamination of sensitive processes or equipment.

16. **Medical Assessment:**

Individuals reporting health issues may be required to undergo a medical assessment. This can help determine the nature of the illness and assess whether the individual is fit to return to work in the cleanroom.

17. **Return-to-Work Clearance:**

Clear guidelines should be established for when individuals can return to work in the cleanroom after experiencing a health issue. Medical clearance may be required to ensure that the individual is no longer a risk to the controlled environment.

18. **Education and Training:**

Conduct regular education and training sessions on the importance of reporting health issues. This includes providing information on common symptoms, the significance of early reporting, and the collective responsibility of all cleanroom users in maintaining a healthy environment.

19. **Hygiene Practices:**

Reinforce hygiene practices as a preventive measure. This includes proper handwashing, the use of personal protective equipment (PPE), and adherence to cleanroom gowning procedures to minimize the risk of spreading illnesses.

20. **Monitoring and Documentation:**

Establish a system for monitoring reported health issues and documenting the actions taken. This documentation contributes to ongoing improvements in health and safety protocols within the cleanroom.

By fostering a culture of prompt reporting and addressing health issues, cleanroom facilities can safeguard both personnel and the integrity of critical processes. These measures contribute to the overall success of cleanroom operations and the reliability of sensitive equipment within the controlled environment.

## 2.6 How to Review the Cleanroom Safety Procedure?

nsuring that all safety procedures are followed for the safe operation of a cleanroom involves a comprehensive approach that encompasses training, monitoring, and continuous improvement. Here's a detailed guide on how to achieve this:

**1. Safety Training:**

- **Initial Orientation:**

    - Conduct thorough safety orientations for all personnel entering the cleanroom.

    - Cover basic safety rules, emergency procedures, and the use of personal protective equipment (PPE).

    - Ensure that individuals are aware of cleanroom classifications, restricted behaviors, and specific safety protocols.

- **Regular Training Sessions:**

    - Schedule regular safety training sessions, including refresher courses, to reinforce safety protocols.

    - Address any updates or changes to safety procedures promptly.

**2. Documentation and Communication:**

- **Standard Operating Procedures (SOPs):**

    - Develop and maintain comprehensive SOPs outlining safety procedures in the cleanroom.

    - Clearly document step-by-step instructions for handling hazardous materials, emergency response, and evacuation plans.

- **Visible Signage:**

    - Install visible and easily understandable signage indicating safety rules and emergency procedures.

    - Ensure that signs are strategically placed at entry points, exits, and within the cleanroom.

**3. Safety Equipment and Infrastructure:**

- **Availability of Safety Equipment:**
    - Ensure that necessary safety equipment, such as eye wash stations, emergency showers, fire extinguishers, and first aid kits, is readily available.
    - Regularly inspect and maintain safety equipment to ensure functionality.
- **Emergency Exits and Lighting:**
    - Clearly mark emergency exits and ensure that they are unobstructed.
    - Maintain proper lighting, especially in critical areas, to facilitate safe movement.

**4. Personnel Accountability:**

- **Access Control:**
    - Implement access control measures to restrict entry to authorized personnel only.
    - Regularly update access permissions and conduct audits to ensure compliance.
- **Accountability Checks:**
    - Conduct periodic accountability checks to verify that all individuals present in the cleanroom are authorized and trained.
    - Use electronic systems or logbooks to track personnel entry and exit times.

**5. Monitoring and Auditing:**

- **Regular Inspections:**
    - Schedule routine inspections of the cleanroom environment to identify any potential safety hazards.
    - Inspect equipment, infrastructure, and cleanliness levels regularly.
- **Internal Audits:**
    - Conduct internal safety audits to assess adherence to safety procedures.
    - Use checklists and established criteria to evaluate safety compliance.

## 6. Emergency Response Drills:

- **Regular Drills:**
    - Organize regular emergency response drills, including fire drills, chemical spill response, and evacuation exercises.
    - Evaluate the effectiveness of response procedures and address any shortcomings.

## 7. Continuous Improvement:

- **Feedback Mechanism:**
    - Establish a feedback mechanism for cleanroom personnel to report safety concerns or suggest improvements.
    - Encourage a culture of continuous improvement and learning.
- **Incident Investigation:**
    - Conduct thorough investigations into any safety incidents or near misses.
    - Implement corrective actions to prevent the recurrence of similar incidents.

## 8. Regulatory Compliance:

- **Stay Informed:**
    - Stay updated on relevant safety regulations and standards applicable to cleanroom operations.
    - Ensure that the cleanroom facility is in compliance with local, national, and industry-specific safety guidelines.

## Conclusion:

A systematic and proactive approach to safety, combined with ongoing training and monitoring, is essential for maintaining a safe cleanroom environment. Regular reviews, audits, and feedback loops contribute to continuous improvement and the prevention of safety incidents. It's crucial to foster a safety culture where all personnel understand the importance of adhering to established safety procedures for the well-being of themselves and others.

# Chapter 3 Gowning Procedures Orientation

## 3.1 Proper cleanroom attire

Proper cleanroom attire refers to the specific clothing and accessories worn by individuals working in cleanroom environments to minimize the introduction and generation of contaminants. The attire is designed to maintain the cleanliness and integrity of the controlled environment, protecting sensitive processes, equipment, and products from contamination.

Key elements of proper cleanroom attire include:

1. **Coveralls or Cleanroom Garments:**

   Individuals typically wear one-piece coveralls made from cleanroom-compatible materials. The design and construction of these coveralls aim to minimize particle shedding and prevent the release of contaminants from clothing.

2. **Hoods or Head Coverings:**

   Hoods are worn to cover the hair completely. This prevents hair and skin particles from being introduced into the cleanroom environment. Hoods are an integral part of the attire, particularly in environments with stringent cleanliness requirements.

3. **Boots or Shoe Covers:**

   Cleanroom-compatible boots or shoe covers are worn to prevent contaminants from the shoes from entering the cleanroom. Boots may cover the entire foot and lower leg, providing a barrier against particle release.

4. **Gloves:**

   Cleanroom-compatible gloves are an essential part of attire to cover the hands and wrists. The choice of gloves depends on the specific requirements of the cleanroom, considering factors such as material compatibility, cleanliness, and the level of dexterity needed for tasks.

5. **Face Masks or Respirators:**

   In some cleanroom environments, individuals may be required to wear face masks or respirators to protect against airborne contaminants. The type of respiratory protection depends on the cleanroom classification and the nature of the work being conducted.

6. **Cleanroom Goggles or Safety Glasses:**

> Eye protection, such as cleanroom goggles or safety glasses, may be required to prevent the entry of particles into the eyes. This is especially important in environments where eye protection is necessary for safety and cleanliness.

The goal of proper cleanroom attire is to create a barrier between individuals and the cleanroom environment, reducing the risk of introducing contaminants. The attire is carefully selected and maintained to meet the cleanliness standards of the specific cleanroom classification. Adherence to gowning procedures and protocols ensures that personnel contribute to maintaining the required cleanliness levels within the cleanroom.

## 3.2 Dressing and undressing protocols

Gowning in a cleanroom involves a set of dressing and undressing protocols to ensure that individuals entering the controlled environment do not introduce contaminants. The steps and sequence of gowning are designed to minimize the risk of particle generation and maintain the cleanliness of the cleanroom. Here is a typical sequence for gowning in a cleanroom:

## 3.2.1 Dressing Protocol:

1. **Wash and Dry Hands:**

> Before gowning, individuals must thoroughly wash their hands using a suitable antimicrobial soap. Hands should be dried using lint-free, disposable towels.

2. **Don Cleanroom Undergarments:**

> Individuals typically start by wearing cleanroom undergarments, including undergarment coveralls or other appropriate attire.

3. **Don Shoe Covers or Boots:**

> Cleanroom-compatible shoe covers or boots are worn to prevent contaminants from shoes from entering the cleanroom.

4. **Wear a Hairnet or Hood:**

> Individuals cover their hair completely using a hairnet or hood. This step is crucial in preventing the release of hair and skin particles.

5. **Don Cleanroom Coveralls or Garments:**

   One-piece coveralls or cleanroom garments are worn to cover the entire body. These garments are designed to minimize particle shedding and protect against the release of contaminants.

6. **Put on Gloves:**

   Cleanroom-compatible gloves are worn to cover the hands and wrists. The choice of gloves depends on the specific requirements of the cleanroom and the tasks being performed.

7. **Wear Face Mask or Respirator:**

   If required by cleanroom protocols, individuals put on a face mask or respirator to protect against airborne contaminants.

8. **Don Cleanroom Goggles or Safety Glasses:**

   Eye protection, such as cleanroom goggles or safety glasses, is worn to prevent the entry of particles into the eyes.

## 3.2.2 Undressing Protocol:

1. **Enter the Anteroom or Transition Area:**

   Before leaving the cleanroom, individuals typically enter an anteroom or transition area. This area serves as a buffer zone between the cleanroom and external environments.

2. **Remove Contaminated Items:**

   In the anteroom, individuals remove contaminated items, such as gloves, goggles, and face masks, and dispose of them properly.

3. **Undress in Designated Order:**

   Gowning should be reversed in a designated order to prevent cross-contamination. For example, individuals may start by removing coveralls, then shoe covers, followed by hairnets or hoods.

4. **Dispose of Contaminated Items:**

   All disposed items, including gloves and other disposable protective gear, should be placed in designated waste bins or containers.

5. **Wash and Dry Hands:**

> After undressing, individuals wash their hands thoroughly using antimicrobial soap and dry them with lint-free, disposable towels.

The justification for these steps and sequence lies in maintaining a unidirectional flow of gowning and undressing to prevent the spread of contaminants. By following a systematic and controlled process, individuals contribute to the overall cleanliness and integrity of the cleanroom environment.

## 3.3 Personal hygiene practices

Personal hygiene practices play a crucial role in maintaining the cleanliness and integrity of a cleanroom environment. Adhering to proper personal hygiene practices is essential during the gowning process to prevent the introduction of contaminants. Here are key aspects of personal hygiene practices in cleanroom gowning:

1. **Hand Washing:**

> Before entering the cleanroom, individuals must thoroughly wash their hands using an approved antiseptic soap or disinfectant. This step is critical for removing dirt, oils, and microorganisms from the hands.

2. **Nail and Skin Care:**

> Individuals should keep their nails short and clean to prevent the accumulation of dirt and microorganisms. Long or artificial nails can harbor contaminants and are generally not permitted in cleanroom environments.

3. **No Makeup and Minimal Use of Personal Care Products:**

> Makeup, lotions, and other personal care products can release particles into the air. Cleanroom gowning protocols often restrict or limit the use of such products to minimize the risk of contamination.

4. **Avoidance of Scented Products:**

> Scented products, such as perfumes and colognes, are typically discouraged in cleanrooms. These products may contain volatile organic compounds (VOCs) that can contribute to contamination.

5. **No Eating, Drinking, or Smoking:**

> Individuals should refrain from eating, drinking, or smoking in or around cleanroom areas. These activities can introduce particles, odors, and contaminants into the environment.

6. **Health Status Disclosure:**

   Individuals must disclose any health conditions, illnesses, or recent exposure to contagious diseases before entering the cleanroom. This helps prevent the spread of illnesses and ensures that personnel with potential health risks take appropriate precautions.

7. **Uniform and Work Clothing:**

   Cleanroom-specific uniforms or work clothing should be worn as part of the gowning process. These garments are designed to minimize particle release and maintain a clean environment.

8. **Hair Control:**

   Proper hair control measures, such as wearing a hairnet or a head cover, are essential. Loose hair or dandruff can contribute to airborne particles and should be prevented.

Adhering to these personal hygiene practices ensures that individuals entering cleanroom environments are as clean as possible, reducing the risk of contamination. The sequence and justification of these steps are designed to create a comprehensive approach to maintaining cleanliness and meeting the stringent requirements of the cleanroom classification. The goal is to minimize the introduction of particles and microorganisms, supporting the integrity of sensitive processes conducted within the cleanroom.

# Chapter 4. Cleanroom Etiquette Orientation

## 4.1 Restricted behaviors

Cleanroom etiquette plays a pivotal role in maintaining the cleanliness and integrity of controlled environments. Establishing a set of restricted behaviors is essential to prevent contamination and ensure compliance with cleanroom protocols. Here is an explanation of restricted behaviors during a Cleanroom Etiquette Orientation:

1. **No Eating, Drinking, or Smoking:**

   One of the fundamental rules in cleanroom environments is the prohibition of eating, drinking, or smoking. Consuming food or beverages in the cleanroom introduces the risk of particle and microbial contamination. Smoking is not only a source of particles but also poses a fire hazard.

2. **Avoiding Touching the Face:**

   Touching the face, especially the mouth and nose, can transfer particles from hands to the face. Cleanroom personnel are advised to refrain from touching their faces to minimize the risk of introducing contaminants.

3. **Restricted Talking and Loud Noises:**

   Excessive talking can generate particles through speech, and loud noises may disrupt sensitive processes. Cleanroom etiquette often encourages personnel to maintain a quiet environment to reduce particle generation and ensure focus on critical tasks.

4. **No Chewing Gum:**

   Chewing gum is discouraged in cleanrooms as it can lead to particle release. Additionally, gum can be a source of contamination if accidentally dropped or improperly disposed of.

5. **Limiting Personal Items:**

   Bringing personal items into the cleanroom should be minimized. Items such as pens, notebooks, or other non-essential personal belongings may carry contaminants. Cleanroom etiquette often involves leaving unnecessary items outside the cleanroom or placing them in designated areas.

6. **Avoiding Unnecessary Movements:**

   Excessive movements can stir up particles from clothing and accessories. Cleanroom personnel are trained to move deliberately and with purpose to reduce the potential for particle release.

7. **No Entry Without Proper Gowning:**

   Entry into the cleanroom is strictly prohibited without proper gowning. This includes wearing the prescribed cleanroom attire, following gowning procedures, and undergoing necessary cleanliness checks.

8. **Reporting Accidents or Contamination Immediately:**

   If an accidental spill, contamination, or any unexpected incident occurs, personnel are required to report it immediately. Prompt reporting ensures timely intervention and minimizes the impact on cleanroom processes.

These restricted behaviors are integral to cleanroom etiquette, emphasizing the importance of individual responsibility in maintaining a contaminant-free environment. Cleanroom personnel undergo thorough orientation to understand and adhere to these guidelines, contributing to the overall success of cleanroom operations and the protection of sensitive processes.

## 4.2 Movement and traffic flow guidelines:

Cleanroom etiquette encompasses specific guidelines for movement and traffic flow to maintain cleanliness, prevent contamination, and ensure the optimal functioning of controlled environments. During a Cleanroom Etiquette Orientation, individuals are educated on the following movement and traffic flow guidelines:

1. **Designated Entry and Exit Points:**

   Cleanrooms have designated entry and exit points equipped with air showers or gowning areas. Individuals must use these specified points for entering and exiting to control the flow of particles.

2. **Unidirectional Flow:**

   Cleanrooms often implement unidirectional flow, where personnel move in a single direction to minimize the potential for cross-contamination. This flow is designed based on the cleanroom's layout and the location of critical processes.

3. **Adherence to Cleanroom Classes:**

   Cleanrooms are classified based on their cleanliness levels, defined by the number of allowable particles per cubic meter. Personnel should be aware of the cleanroom class they are entering and adhere to the corresponding cleanliness standards.

4. **Minimizing Traffic:**

   Cleanroom personnel are encouraged to minimize unnecessary movements. Excessive traffic can increase the introduction of particles into the environment. Individuals should only move within the cleanroom if it is essential to their assigned tasks.

5. **Avoiding Congestion:**

   Congestion in certain areas can disrupt laminar flow systems and compromise cleanliness. Cleanroom etiquette emphasizes the avoidance of congestion by regulating the number of personnel present in specific zones.

6. **Scheduled Work Times:**

   Cleanroom activities, especially those requiring significant movements, are often scheduled to minimize disruptions. This includes planned maintenance, equipment adjustments, and other tasks that may necessitate increased personnel presence.

7. **Handling of Materials:**

   Movement of materials into and within the cleanroom follows established protocols. Material transfer points are designated, and the movement of supplies is carefully controlled to prevent the introduction of contaminants.

8. **Awareness of Surroundings:**

   Cleanroom personnel are trained to be aware of their surroundings and the activities of others. This awareness helps prevent collisions, spills, or other incidents that could lead to contamination.

9. **Emergency Procedures:**

   Cleanroom etiquette includes knowledge of emergency procedures and evacuation routes. Personnel must understand how to exit the cleanroom quickly and safely in case of emergencies.

10. **Use of Pass-Through Chambers:**

Some cleanrooms feature pass-through chambers for transferring materials between controlled environments. Personnel are instructed on the proper use of these chambers to maintain cleanliness.

By adhering to these movement and traffic flow guidelines, individuals contribute to the overall success of cleanroom operations. These protocols are essential for preserving the integrity of processes and ensuring the reliability of products manufactured within cleanroom environments.

## 4.3 Communication protocols

Communication is a crucial aspect of maintaining a cleanroom environment, where precision, collaboration, and minimizing contamination are paramount. Cleanroom etiquette includes specific communication protocols to ensure effective information exchange while upholding cleanliness standards. During a Cleanroom Etiquette Orientation, individuals are educated on the following communication guidelines:

1. **Use of Cleanroom-Approved Communication Devices:**

   Cleanrooms often have restrictions on the types of electronic devices allowed. Cleanroom personnel are instructed to use communication devices that meet cleanliness standards, such as specially designed headsets or devices housed in cleanroom-compatible enclosures.

2. **Designated Communication Areas:**

   - Specific zones within or adjacent to the cleanroom may be designated for communication activities. These areas are equipped with cleanroom-compatible communication tools to facilitate discussions without compromising cleanliness.

3. **Minimized Verbal Communication:**

   Excessive talking can introduce particles into the cleanroom environment. Cleanroom etiquette encourages personnel to minimize verbal communication, especially in critical areas, and to use alternative means of conveying information when possible.

4. **Clear and Concise Communication:**

   When communication is necessary, it should be clear, concise, and focused on the task at hand. Detailed instructions, warnings, or updates should be communicated with precision to avoid misunderstandings.

5. **Written Communication:**

In situations where verbal communication is challenging or restricted, written communication may be employed. This can include the use of cleanroom-compatible notebooks, whiteboards, or electronic displays to convey important information.

6. **Use of Visual Signals:**

Visual signals or cues may be established for specific communication needs. For example, a designated signal could indicate the need for assistance, a process issue, or an emergency, allowing personnel to convey information without speaking.

7. **Training on Cleanroom-Safe Communication Practices:**

Cleanroom personnel receive training on cleanroom-safe communication practices. This includes understanding the potential impact of speech, gestures, or devices on cleanliness and taking precautions to mitigate these risks.

8. **Emergency Communication Procedures:**

Protocols for emergency communication are established, including the use of alarms, signals, or designated communication channels. Cleanroom personnel are familiarized with these procedures to respond promptly to any urgent situations.

9. **Respect for Quiet Zones:**

Certain areas within the cleanroom may be designated as "quiet zones" where noise levels are minimized. Cleanroom etiquette emphasizes the need for personnel to respect these zones to prevent disturbances that could compromise processes.

By adhering to these communication protocols, cleanroom personnel contribute to maintaining a controlled and efficient environment. These guidelines help prevent contamination, enhance collaboration, and ensure that critical information is conveyed effectively within the cleanroom setting.

# Chapter 5. Tool and Equipment Usage Orientation

## 5.1 Operation guidelines

During a Cleanroom Tool and Equipment Usage Orientation, individuals are provided with comprehensive guidelines on the proper operation of tools and equipment within the cleanroom environment. Adhering to these guidelines is essential to ensure the integrity of processes, prevent contamination, and maintain a controlled cleanroom setting. The operation guidelines include:

1. **Authorized Personnel Only:**

    Access to cleanroom tools and equipment is restricted to authorized personnel with the necessary training and qualifications. Unauthorized individuals are prohibited from operating or adjusting any tools.

2. **Understanding Equipment Functions:**

    Cleanroom personnel are required to have a thorough understanding of the functions and specifications of the tools and equipment they operate. This knowledge ensures proper usage and helps prevent misuse or errors.

3. **Pre-Use Inspections:**

    Prior to operation, individuals must conduct pre-use inspections of tools and equipment. This includes checking for any visible damage, ensuring cleanliness, and verifying that all components are in working order.

4. **Adherence to Standard Operating Procedures (SOPs):**

    Cleanroom tools and equipment are operated in accordance with established Standard Operating Procedures (SOPs). Personnel receive training on these procedures and are expected to follow them meticulously to maintain consistency and reliability in processes.

5. **Calibration and Maintenance:**

    Tools and equipment requiring calibration undergo regular calibration checks to ensure accuracy. Additionally, routine maintenance schedules are followed to address wear and tear, with any necessary repairs performed promptly.

6. **Proper Start-Up and Shutdown Procedures:**

   Cleanroom personnel are educated on the correct start-up and shutdown procedures for each tool or piece of equipment. This includes following a specific sequence to minimize the risk of contamination and ensure optimal performance.

7. **Use of Cleanroom-Compatible Consumables:**

   Consumables, such as wipes, gloves, or lubricants, used in conjunction with tools and equipment must be cleanroom-compatible. Personnel are trained to select and apply consumables that meet cleanliness standards.

8. **Monitoring and Recording:**

   Continuous monitoring of tool performance is conducted during operation. Cleanroom personnel are responsible for recording relevant data and observations, contributing to the maintenance of comprehensive records for quality control purposes.

9. **Emergency Procedures:**

   Cleanroom personnel are trained to respond to equipment malfunctions or emergencies promptly. This includes knowledge of emergency shutdown procedures and the use of safety features to minimize risks.

10. **Reporting Issues:**

    Any issues, malfunctions, or deviations from normal operation are promptly reported to supervisors or relevant personnel. This proactive reporting ensures timely intervention and prevents potential disruptions to cleanroom processes.

11. **Tool-Specific Training:**

    Personnel receive specific training on each tool or equipment they operate. This training covers not only the technical aspects of operation but also emphasizes the importance of cleanliness and contamination control.

By following these operation guidelines, cleanroom personnel contribute to maintaining the cleanliness, efficiency, and reliability of tools and equipment, ultimately upholding the stringent standards of the cleanroom environment.

## 5.2 Proper handling of tools and equipment

As part of the Cleanroom Tool and Equipment Usage Orientation, individuals receive guidance on the proper handling of tools and equipment within the cleanroom environment. Proper handling is crucial to prevent contamination, maintain equipment integrity, and ensure the overall cleanliness and functionality of the cleanroom. The guidelines for the proper handling of tools and equipment include:

1. **Gentle Handling:**

   Cleanroom personnel are instructed to handle tools and equipment with care, avoiding unnecessary force or impact. This gentle handling helps prevent damage to sensitive components and minimizes the generation of particles.

2. **Minimized Contact:**

   Individuals are trained to minimize direct contact with critical surfaces of tools and equipment. Touching surfaces with bare hands can transfer oils, skin particles, or contaminants, compromising the cleanliness of the equipment.

3. **Use of Cleanroom-Compatible Gloves:**

   When direct contact is necessary, personnel must wear cleanroom-compatible gloves. These gloves act as a barrier, preventing direct skin contact and reducing the risk of transferring contaminants to the tools or equipment.

4. **Prohibited Materials:**

   Cleanroom users are informed about materials that should not come into contact with tools and equipment. This includes substances that may corrode or damage surfaces, as well as materials that could introduce contaminants.

5. **Proper Lifting Techniques:**

   Training includes proper lifting techniques when handling heavy or delicate equipment. Incorrect lifting can lead to damage, and proper techniques help prevent accidents and maintain the integrity of the tools.

6. **Controlled Movement:**

   Cleanroom personnel are advised to move tools and equipment in a controlled manner to avoid collisions or sudden movements that could dislodge particles. Smooth and deliberate movements contribute to a controlled cleanroom environment.

7. **Utilizing Handling Aids:**

   When applicable, handling aids such as carts, carriers, or other designated tools are provided and should be used for transporting equipment. This minimizes direct handling and reduces the risk of contamination.

8. **Adherence to Equipment Labels:**

   All instructions and labels on tools and equipment are to be carefully read and followed. This includes any specific handling instructions or precautions outlined by the manufacturer.

9. **Avoiding Unnecessary Adjustments:**

   Cleanroom users are discouraged from making unnecessary adjustments or modifications to tools and equipment. Any modifications should be performed by qualified personnel following approved procedures.

10. **Return to Designated Locations:**

    After use, tools and equipment should be returned to their designated locations promptly. This ensures that they are stored in controlled environments, minimizing the risk of contamination during storage.

11. **Tool-Specific Handling Instructions:**

    Personnel receive specific instructions for the handling of each tool or piece of equipment they use. These instructions address unique considerations and requirements for individual tools.

By adhering to these guidelines for proper handling, cleanroom personnel contribute to maintaining the cleanliness, functionality, and reliability of tools and equipment within the controlled cleanroom environment.

# Chapter 6. Contamination Control Orientation

## 6.1 Understanding sources of contamination

Contamination control is a critical aspect of cleanroom management, and individuals undergoing orientation learn to understand the various sources of contamination within a cleanroom environment. This knowledge is essential for preventing and mitigating contamination risks. The orientation on understanding sources of contamination includes:

1. **Airborne Particles:**

    Cleanroom users are educated about the presence of airborne particles, including dust, lint, and other microscopic debris. These particles can originate from various sources, such as clothing, equipment, or external environments, and can compromise the cleanroom's controlled atmosphere.

2. **Personnel:**

    Human activities are a significant source of contamination. Cleanroom personnel are trained to be aware of shedding skin cells, hair, and particles from clothing. Proper gowning procedures and personal hygiene practices are emphasized to minimize the introduction of contaminants by individuals.

3. **Tools and Equipment:**

    Tools and equipment, if not handled properly, can introduce particles or residues. The orientation covers the proper handling, cleaning, and maintenance of tools and equipment to prevent them from becoming sources of contamination.

4. **Materials and Supplies:**

    Materials brought into the cleanroom, including chemicals, substrates, or consumables, can be potential sources of contamination. Cleanroom users are instructed on the proper inspection, storage, and handling of materials to avoid introducing contaminants.

5. **Ventilation Systems:**

    The cleanroom's ventilation system plays a crucial role in controlling airborne particles. Individuals learn how the ventilation system works, the importance of proper air exchange rates, and the need to prevent blockages or malfunctions that could compromise air quality.

6. **Entry and Exit Points:**

The entry and exit points of a cleanroom are sensitive areas where contaminants can be introduced. Users are trained on proper procedures for entering and exiting, including gowning and ungowning protocols, to minimize the risk of contamination.

7. **Maintenance Activities:**

Maintenance and cleaning activities, if not conducted carefully, can lead to the release of particles. Cleanroom personnel receive guidance on performing maintenance tasks without compromising the cleanroom environment.

8. **Utilities and Services:**

Utilities such as water, gases, and other services entering the cleanroom must meet specified cleanliness standards. Orientation includes information on monitoring and maintaining the quality of utilities to prevent contamination.

9. **External Environment:**

Cleanroom users understand the potential impact of the external environment on contamination. They are informed about procedures for mitigating contamination risks when bringing items into the cleanroom from outside.

10. **Waste Disposal:**

Proper waste disposal practices are crucial for preventing contaminants from accumulating within the cleanroom. Cleanroom personnel learn about segregation, labeling, and disposal procedures for different types of waste.

11. **Equipment Exhaust:**

Certain equipment may release exhaust that can contain particles or fumes. Individuals are trained to ensure that equipment exhaust is properly managed to prevent contamination of the cleanroom air.

By comprehensively understanding these sources of contamination, cleanroom users are better equipped to adhere to protocols and practices that contribute to maintaining a controlled and contaminant-free environment.

## 6.2 Mitigation strategies

Contamination mitigation is a critical aspect of cleanroom management to maintain a controlled environment with minimal pollutants. Cleanroom personnel undergo orientation to learn various strategies and practices aimed at mitigating contamination risks. The contamination control orientation includes:

**Gowning Procedures:**

Proper gowning is one of the primary strategies to prevent the introduction of contaminants. Cleanroom users are trained on the correct procedures for donning and doffing cleanroom garments, including coveralls, hoods, gloves, and shoe covers.

**Personnel Hygiene:**

Emphasis is placed on personal hygiene practices to minimize the shedding of skin particles, hair, and other contaminants. This includes thorough handwashing, use of approved personal care products, and adherence to cleanliness standards.

**Cleanroom Behavior and Etiquette:**

Cleanroom users are educated on appropriate behaviors within the controlled environment. This includes movement protocols, restricted activities (e.g., no eating or drinking), and adherence to communication guidelines to prevent the generation of particles.

**Tool and Equipment Handling:**

Proper handling of tools and equipment is crucial to prevent the release of particles. Cleanroom personnel receive training on the correct use, cleaning, and maintenance of tools to minimize the risk of introducing contaminants.

**Material Handling:**

Careful handling of materials and supplies is emphasized to prevent contamination. This includes proper inspection, storage, and transportation procedures to ensure that materials entering the cleanroom meet cleanliness standards.

**Controlled Environment Monitoring:**

Regular monitoring of the cleanroom environment is essential. Cleanroom users learn about the monitoring systems in place, including particle counters, temperature and humidity sensors, and how to respond to deviations from specified cleanliness levels.

**Airflow Control:**

Understanding the cleanroom's airflow patterns and maintaining proper air exchange rates is crucial for controlling airborne contaminants. Cleanroom personnel are trained to avoid disrupting airflow and to report any issues promptly.

**Equipment Maintenance:**

Regular maintenance of cleanroom equipment is essential to prevent malfunctions that could lead to contamination. Personnel learn about the importance of scheduled maintenance and the procedures for conducting equipment checks.

**Waste Management:**

Proper waste disposal practices are integral to contamination control. Cleanroom users are educated on the correct segregation, labeling, and disposal procedures for different types of waste generated within the cleanroom.

**Training and Education:**

Ongoing training and education programs are implemented to keep cleanroom personnel updated on best practices and new contamination control measures. This ensures a continuous understanding of the importance of contamination mitigation.

**Emergency Response Planning:**

Cleanroom users are briefed on emergency response plans in case of unexpected events. Having a well-defined plan helps mitigate the risk of contamination during emergencies.

By incorporating these mitigation strategies into their daily practices, cleanroom personnel contribute to the overall effectiveness of contamination control, maintaining a pristine and controlled environment conducive to sensitive processes and manufacturing.

## 6.3 Cleanroom cleaning protocols:

Cleanroom cleaning protocols are essential to maintaining a controlled environment with minimal contaminants. These protocols are designed to ensure that surfaces, equipment, and the overall cleanroom environment meet specified cleanliness standards. Cleanroom personnel are trained on the following cleaning protocols:

### Cleaning Frequency:

Establish a regular cleaning schedule based on the cleanroom's classification and usage. High-traffic or critical areas may require more frequent cleaning to maintain cleanliness levels.

### Cleaning Agents and Disinfectants:

Identify and use approved cleaning agents and disinfectants suitable for the cleanroom environment. These substances should effectively remove or neutralize contaminants without leaving residues that could interfere with processes.

### Surface Cleaning:

Instruct cleanroom personnel on proper surface cleaning techniques. This involves wiping surfaces methodically using lint-free, non-shedding wipes or mop heads soaked in the appropriate cleaning solution.

### Workstation and Equipment Cleaning:

Emphasize the importance of cleaning workstations, tools, and equipment regularly. Special attention should be given to areas where particles can accumulate, such as seams, joints, and crevices.

### Horizontal and Vertical Surfaces:

Train personnel to clean both horizontal and vertical surfaces systematically. This includes walls, ceilings, countertops, and equipment surfaces to prevent the settling of particles.

### Floor Cleaning:

Provide guidelines on cleaning cleanroom floors, which may involve specific mopping techniques and the use of specialized cleaning agents. Contaminants on the floor can become airborne, affecting the overall cleanliness of the cleanroom.

### Gowning Room and Anteroom Cleaning:

Cleanroom gowning rooms and anterooms are critical transitional spaces. Ensure that these areas are regularly cleaned to prevent the transfer of contaminants from outside into the cleanroom.

### Cleaning Validation:

Implement procedures for cleaning validation to verify the effectiveness of cleaning processes. This may involve periodic testing and monitoring to ensure that cleanliness levels are consistently met.

### Documentation:

Stress the importance of documentation related to cleaning activities. Cleanroom personnel should maintain records of cleaning schedules, activities performed, and any deviations observed during the cleaning process.

### Training on Cleanroom Cleaning Equipment:

Provide training on the proper use of cleaning equipment, including mop systems, vacuum cleaners with HEPA filters, and any specialized tools designed for cleanroom cleaning. Incorrect use of equipment can contribute to contamination.

### Specialized Cleaning for Critical Environments:

In environments with stringent cleanliness requirements, such as semiconductor cleanrooms, provide specialized training for critical cleaning procedures. This may involve more rigorous protocols and specific cleaning agents.

### Adherence to Cleanroom Etiquette:

Reinforce the link between personal behaviors and cleanliness. Emphasize that maintaining personal cleanliness, proper gowning, and following cleanroom etiquette contribute to overall cleanliness.

By incorporating these cleanroom cleaning protocols into their routine activities, cleanroom personnel play a vital role in ensuring that the controlled environment remains free from contaminants, meeting the stringent requirements of various industries and processes.

# Chapter 7. Waste Disposal Procedures

## 7.1 Segregation practices (e.g., separating general waste from hazardous waste)

Segregation practices in cleanroom waste disposal are crucial to prevent cross-contamination, maintain cleanliness, and adhere to environmental regulations. Cleanroom personnel must follow specific guidelines for segregating and disposing of different types of waste. Here are key points regarding segregation practices:

1. **Waste Identification:**

   Clearly identify and categorize different types of waste generated in the cleanroom. This includes general waste, hazardous waste, recyclables, and any other specific categories based on cleanroom activities.

2. **Color-Coded Bins:**

   Implement a color-coded bin system to facilitate easy identification and segregation of waste. Assign distinct colors to bins for general waste, hazardous waste, and recyclables. This helps minimize errors in disposal.

**Proper Labeling:**

   Ensure that each waste bin is properly labeled with the type of waste it is intended for. Clear and standardized labels aid in quick identification and prevent confusion among cleanroom personnel.

**Hazardous Waste Separation:**

   Clearly define the procedures for handling and segregating hazardous waste. Hazardous materials must be segregated from general waste to prevent contamination and comply with regulations governing the disposal of hazardous substances.

**Recyclable Materials:**

   Provide separate bins for recyclable materials such as paper, plastics, and glass. Recycling practices contribute to environmental sustainability and reduce the overall waste generated in the cleanroom.

**Segregation at Source:**

   Encourage cleanroom personnel to segregate waste at the source, i.e., where it is generated. This helps minimize the mixing of different waste streams and streamlines the disposal process.

3. **Training and Awareness:**

> Conduct regular training sessions to educate cleanroom personnel about the importance of waste segregation and the specific procedures to follow. Increased awareness helps prevent accidental mixing of incompatible waste.

**Adherence to Regulations:**

> Ensure that waste segregation practices align with local, regional, and national regulations governing waste disposal. Compliance with regulations is essential for environmental responsibility and legal adherence.

**Segregation in Gowning and Ancillary Areas:**

> Extend waste segregation practices to gowning rooms and ancillary areas connected to the cleanroom. Waste generated in these areas should also be properly segregated before entering the cleanroom environment.

**Documentation:**

> Maintain detailed records of waste segregation activities. Document the types and quantities of waste generated, the disposal methods employed, and any deviations from standard procedures.

**Regular Audits and Inspections:**

> Conduct regular audits and inspections of the waste disposal process to ensure compliance with segregation practices. Identify and rectify any issues promptly to maintain a high standard of cleanliness.

**Continuous Improvement:**

> Encourage feedback from cleanroom personnel regarding waste disposal procedures. Use this feedback to implement continuous improvement initiatives, refining segregation practices for enhanced efficiency and effectiveness.

By incorporating these segregation practices into cleanroom waste disposal procedures, organizations can ensure a systematic and compliant approach to managing waste while maintaining the integrity of the controlled environment.

## 7.2 Proper disposal of materials

Proper disposal of materials in cleanroom waste management is essential to maintain a contamination-free environment and comply with environmental regulations. Here are key considerations for the proper disposal of materials:

1. **Segregation at Source:**

   Ensure that cleanroom personnel segregate waste at the source based on predefined categories such as general waste, hazardous waste, and recyclables. This initial separation facilitates proper disposal procedures.

**Containerization:**

Use appropriate containers for different types of waste materials. Containers should be leak-proof, sturdy, and properly labeled to indicate the type of waste they contain. This ensures safe handling during disposal.

**Hazardous Material Disposal:**

Establish clear protocols for the disposal of hazardous materials. Hazardous waste must be handled, stored, and transported in accordance with local, regional, and national regulations. Utilize specialized services for the disposal of hazardous substances.

**Recyclable Materials:**

Implement recycling programs for materials that can be recycled, such as paper, plastics, and glass. Clearly mark recycling bins and provide guidance on the proper disposal of recyclable materials to promote environmental sustainability.

**Incineration and Autoclaving:**

Certain types of waste, especially biological and medical waste, may require incineration or autoclaving to ensure complete sterilization. Establish procedures for the safe disposal of materials through these methods, adhering to regulatory guidelines.

**Chemical Waste Disposal:**

Develop specific procedures for the disposal of chemical waste. This includes proper labeling, packaging, and documentation. Coordinate with licensed waste disposal services to manage chemical waste in accordance with safety and environmental standards.

**Sharp Objects Disposal:**

For cleanrooms involved in medical or laboratory activities, establish protocols for the safe disposal of sharp objects such as needles and blades. Utilize designated puncture-resistant containers to prevent injuries during disposal.

### Radioactive Waste Handling:

If the cleanroom deals with radioactive materials, adhere to strict guidelines for the disposal of radioactive waste. This often involves collaboration with specialized disposal services and regulatory agencies.

### Documentation and Record-Keeping:

Maintain detailed records of all waste disposal activities. Document the types and quantities of materials disposed of, methods used, and any deviations from standard procedures. This documentation is crucial for regulatory compliance and internal audits.

### Training and Awareness:

Conduct regular training sessions to educate cleanroom personnel about the proper disposal of materials. Emphasize the importance of following established procedures to prevent contamination and ensure safety.

### Emergency Response Plan:

Include provisions for emergency situations in the waste disposal plan. Define procedures for handling accidental spills, leaks, or other incidents during the disposal process. Ensure that cleanroom personnel are trained to respond promptly and safely to emergencies.

### Continuous Improvement:

Periodically review and update waste disposal procedures based on feedback, regulatory changes, and emerging best practices. Implement continuous improvement initiatives to enhance the efficiency and safety of material disposal processes.

By emphasizing proper disposal practices, cleanrooms can effectively manage waste, minimize environmental impact, and uphold the stringent cleanliness standards required in controlled environments.

## 7.3 Recycling practices

Implementing recycling practices in cleanroom waste disposal is essential for minimizing environmental impact and promoting sustainability. Here are key considerations for incorporating recycling practices into cleanroom waste disposal procedures:

1. **Material Identification and Segregation:**

   Clearly identify recyclable materials and establish a systematic segregation process. Train cleanroom personnel to differentiate between general waste and recyclables at the source.

2. **Recycling Bins and Containers:**

   Provide designated recycling bins and containers for specific recyclable materials, such as paper, plastics, glass, or metal. Ensure that these containers are easily distinguishable from general waste bins to encourage proper disposal.

3. **Labeling and Color-Coding:**

   Label recycling bins with clear instructions and use color-coding to indicate the types of materials accepted. This visual differentiation aids in quick and accurate disposal by cleanroom personnel.

4. **Educational Initiatives:**

   Conduct regular training sessions to educate cleanroom staff about the importance of recycling and the specific procedures to follow. Create awareness about the environmental benefits of recycling and the role cleanroom personnel play in these efforts.

5. **Recycling Guidelines:**

   Develop and disseminate clear guidelines on how to prepare recyclable materials for disposal. This may include removing contaminants, separating materials by type, and ensuring that items are clean and dry before placing them in recycling bins.

6. **Waste Audit and Monitoring:**

   Conduct periodic waste audits to assess the effectiveness of recycling practices. Monitor the contents of recycling bins to identify any contamination or non-recyclable items. Use this information to refine recycling programs.

7. **Collaboration with Recycling Services:**

   Establish partnerships with reputable recycling services that specialize in processing cleanroom waste. Ensure that these services comply with environmental regulations and adhere to sustainable practices.

8. **Recycling of Laboratory Equipment:**

In cleanrooms involved in laboratory activities, explore options for recycling laboratory equipment and consumables. This may include glassware, plastic containers, and other materials commonly used in scientific research.

9. **Recycling of Packaging Materials:**

Emphasize the recycling of packaging materials received within the cleanroom. Collaborate with suppliers to minimize packaging waste and ensure that any packaging materials are recyclable.

10. **Documentation of Recycling Efforts:**

Maintain records documenting the quantities and types of materials recycled. This documentation serves as evidence of the cleanroom's commitment to environmentally friendly practices and can be useful for regulatory compliance.

11. **Continuous Improvement:**

Regularly assess the effectiveness of recycling practices and seek opportunities for improvement. Stay informed about advancements in recycling technologies and adjust procedures accordingly to enhance the cleanroom's contribution to sustainability.

Incorporating recycling practices into cleanroom waste disposal procedures aligns with broader environmental goals and demonstrates a commitment to responsible waste management within controlled environments.

## 7.4 Disposal Guidelines of Hazardous Materials

## 7.4.1 Disposal Guidelines of Hydrofluoric Acid (HF):

Disposing of hydrofluoric acid (HF) in a cleanroom requires careful handling and adherence to safety and environmental regulations due to the hazardous nature of HF. Here are guidelines for the proper disposal of HF in a cleanroom:

1. **Personal Protective Equipment (PPE):**

Before initiating the disposal process, ensure that personnel involved in handling HF wear appropriate PPE, including gloves, chemical-resistant aprons, face shields, and protective eyewear. This is crucial to prevent direct contact with HF.

2. **Containment and Segregation:**

Contain the HF waste in chemically resistant and properly labeled containers. Segregate HF waste from other types of chemical waste to avoid potential reactions or contamination.

## 3. Neutralization:

Hydrofluoric acid is highly corrosive, and its disposal often involves neutralization. However, neutralizing HF requires specialized procedures. Consult with experienced chemical safety personnel or environmental health and safety experts to determine the appropriate neutralizing agents and methods.

## 4. Dilution:

If neutralization is not feasible, consider diluting the HF waste with a large volume of water to reduce its concentration. Dilution should be performed slowly and carefully to minimize the risk of heat generation or splashing.

## 5. Dedicated HF Waste Container:

Designate a dedicated waste container for HF disposal. Ensure that the container is made of compatible materials resistant to HF and has a secure lid to prevent leakage.

## 6. Labeling:

Clearly label the waste container with information such as "Hydrofluoric Acid Waste," concentration (if known), and any other relevant details. Adhering to labeling requirements ensures that the waste is handled appropriately throughout the disposal process.

## 7. Waste Disposal Service:

Contact a licensed and authorized hazardous waste disposal service to manage the collection, transportation, and treatment or disposal of HF waste. Ensure that the disposal service complies with local, state, and federal regulations.

## 8. Documentation:

Maintain accurate records of the HF waste disposal process, including dates, quantities, and disposal service details. Documentation is essential for regulatory compliance and internal tracking of hazardous waste management.

9. **Emergency Preparedness:**

- Have emergency response procedures in place in case of accidental spills or exposures during the disposal process. This includes providing accessible emergency eyewash stations, showers, and communication protocols for immediate assistance.

10. **Regulatory Compliance:**

Familiarize yourself with local, state, and federal regulations governing the disposal of hazardous materials, including HF. Ensure that all disposal activities align with these regulations to avoid legal and environmental consequences.

It is crucial to emphasize that the disposal of HF should be undertaken by personnel with expertise in handling hazardous chemicals, and all actions should align with established safety protocols and regulatory requirements. If there is uncertainty or lack of experience, seek guidance from chemical safety professionals or environmental health and safety experts.

## 7.4.2 Disposal Guidelines for Piranha Solution:

Piranha solution, a highly reactive mixture of sulfuric acid ($H_2SO_4$) and hydrogen peroxide ($H_2O_2$), is commonly used for cleaning and etching surfaces in cleanroom environments. Due to its strong oxidizing properties, handling and disposal require careful attention to safety and environmental regulations.

**Chemical Formula of Piranha Solution:** The chemical formula for Piranha solution is typically expressed as a mixture of concentrated sulfuric acid and hydrogen peroxide. The reaction between these two compounds generates highly reactive species, including peroxymonosulfuric acid ($HSO_5$) and peroxymonosulfate ions ($HSO_5^-$). The overall reaction can be represented as follows:

$$H_2SO_4 + H_2O_2 \rightarrow HSO_5 + H_2O H_2SO_4 + H_2O_2 \rightarrow HSO_5 + H_2O$$

1. **Neutralization:**

Prior to disposal, consider neutralizing the Piranha solution to reduce its reactivity. This can be achieved by slowly adding a base (such as sodium bicarbonate or sodium hydroxide) while monitoring the pH. Take precautions as the neutralization process may generate heat.

2. **Dilution:**

Dilute the neutralized Piranha solution with a large volume of water. Slowly add water to the solution to minimize the risk of splashing or excessive heat generation. Dilution helps reduce the concentration of reactive species.

3. **Chemical Compatibility:**

   Use containers made of materials that are chemically compatible with the constituents of Piranha solution. Glass or high-density polyethylene (HDPE) containers are often suitable. Avoid using containers that may react with or degrade in the presence of sulfuric acid.

4. **Labeling:**

   Clearly label the container with information indicating the contents, concentration (if known), and any specific handling precautions. Adhering to proper labeling ensures that others are aware of the hazardous nature of the waste.

5. **Professional Disposal Services:**

   - Contact licensed and authorized hazardous waste disposal services to handle the collection, transportation, and treatment or disposal of the neutralized and diluted Piranha solution. These services should comply with local, state, and federal regulations.

6. **Documentation:**

   Maintain detailed records of the neutralization and disposal process, including dates, quantities, and disposal service information. Proper documentation is essential for regulatory compliance and internal tracking.

7. **Emergency Preparedness:**

   Have emergency response procedures in place in case of accidental spills or exposures during the disposal process. This includes providing emergency eyewash stations, showers, and communication protocols for immediate assistance.

It is crucial to follow established safety protocols, regulatory requirements, and guidelines specific to the cleanroom environment when disposing of Piranha solution. If there is uncertainty or lack of experience, seek guidance from chemical safety professionals or environmental health and safety experts.

### 7.4.3 Disposal Guidelines for Wafers in a Cleanroom:

Disposing of wafers in a cleanroom involves several considerations to ensure that it is done in a manner that prioritizes safety, environmental compliance, and adherence to cleanroom protocols. Wafers may be used in various semiconductor and nanotechnology processes and can be contaminated with materials that require proper handling.

1. **Identification of Contaminants:**

   Before disposal, identify any contaminants or residues on the wafers. Different processes may leave behind materials that need specific disposal methods.

2. **Cleaning and Decontamination:**

   Clean the wafers to the best extent possible using approved cleaning procedures within the cleanroom. Decontamination is crucial to minimize the risk of releasing hazardous materials during disposal.

3. **Segregation of Materials:**

   Segregate wafers based on the materials they contain or the processes they underwent. This ensures that similar materials are grouped together for proper disposal methods.

4. **Packaging:**

   Place the cleaned and decontaminated wafers in appropriate packaging that prevents breakage or damage during transportation. Packaging materials should be compatible with cleanroom standards.

5. **Labeling:**

   Clearly label the packaging with information on the contents, potential hazards, and any specific handling instructions. This is essential for the safety of those involved in disposal and for compliance with waste management regulations.

6. **Documentation:**

   Maintain detailed records of the wafers being disposed of, including information on the processes they were involved in, any hazardous materials present, and the cleaning and decontamination procedures performed.

7.  **Hazardous Waste Disposal Services:**

Contact licensed hazardous waste disposal services that specialize in semiconductor and nanotechnology waste. Ensure that the disposal service is compliant with local, state, and federal regulations regarding the disposal of electronic or semiconductor-related waste.

8.  **Recycling Opportunities:**

Investigate whether the wafers or their materials can be recycled. Some cleanroom facilities have specific recycling programs for certain materials. Recycling contributes to sustainability efforts and minimizes the environmental impact of waste disposal.

9.  **Secure Transportation:**

Arrange for secure transportation of the packaged wafers to the disposal facility. Follow guidelines for transporting hazardous materials and ensure that the transportation method is compliant with regulations.

10. **Emergency Preparedness:**

Have emergency response procedures in place in case of accidental spills or releases during the disposal process. This includes providing emergency response equipment and communication protocols for immediate assistance.

It is essential to follow cleanroom protocols, environmental regulations, and safety guidelines when disposing of wafers. If there is uncertainty about the proper disposal procedures, seek guidance from cleanroom management, environmental health and safety professionals, or waste disposal

# Chapter8. Cleanroom Classification and Standards Orientation

## 8.1 Understanding cleanroom classes

Cleanroom classification and standards are essential aspects of maintaining a controlled environment to ensure the quality and integrity of processes conducted within the cleanroom. Cleanrooms are classified based on the level of cleanliness they can maintain, and these classifications adhere to international standards. Understanding cleanroom classes is crucial for all individuals working in or around cleanroom facilities.

**Key Components of Cleanroom Classification:**

1. **Cleanroom Classes:**

   Cleanrooms are classified into different classes, typically ranging from ISO Class 1 to ISO Class 9. The classification is based on the maximum allowable number of airborne particles per cubic meter for particles equal to or larger than a specified size.

2. **ISO Standards:**

   The International Organization for Standardization (ISO) sets the standards for cleanroom classification. ISO 14644-1 is a widely recognized standard that defines the maximum allowable particle counts for each cleanroom class.

3. **Airborne Particles:**

   Cleanroom classes are defined by the concentration of airborne particles of different sizes. The sizes of particles considered are usually 0.5 micrometers and larger, as these can significantly impact sensitive processes.

4. **Particle Counting:**

   Monitoring and controlling airborne particles are critical. Particle counters are used to measure the particle concentration in the air, and the results determine whether a cleanroom complies with its designated class.

5. **Occupancy State:**

   Cleanroom standards also consider the occupancy state, distinguishing between "at rest" (when no personnel are present) and "in operation" (during regular activities). The particle limits may vary depending on the state.

6. **Cleanroom Design and Maintenance:**

    Cleanroom design and maintenance must align with the intended cleanroom class. This includes the selection of appropriate materials, air filtration systems, gowning procedures, and cleaning protocols.

**Cleanroom Classification in Detail:**

- **ISO Class 1 to ISO Class 9:**

    ISO Class 1 represents the cleanest environment with the strictest particle limits, while ISO Class 9 has higher allowable particle counts. The higher the ISO Class, the more relaxed the cleanliness requirements.

**Importance of Understanding Cleanroom Classes:**

1. **Process Integrity:**

    Different processes have specific cleanliness requirements. Understanding cleanroom classes ensures that the cleanroom environment aligns with the needs of the processes conducted within.

2. **Quality Control:**

    Compliance with cleanroom standards is essential for maintaining product quality and preventing contamination in industries such as semiconductor manufacturing, pharmaceuticals, and biotechnology.

3. **Personnel Training:**

    Personnel working in cleanrooms must be aware of the cleanliness standards to adhere to proper gowning procedures, behavior, and hygiene practices.

4. **Facility Management:**

    Facility managers need a thorough understanding of cleanroom classes to design, operate, and maintain cleanroom facilities effectively.

5. **Regulatory Compliance:**

    Many industries are subject to regulatory requirements that mandate adherence to specific cleanroom standards. Understanding these standards ensures compliance.

In summary, a clear understanding of cleanroom classes and standards is fundamental for maintaining the required level of cleanliness, safeguarding sensitive processes, and ensuring regulatory compliance within a cleanroom facility.

## 8.2 Adhering to industry standards

Adhering to industry standards is a critical aspect of maintaining a cleanroom's effectiveness in controlling contamination and ensuring a controlled environment for sensitive processes. Cleanroom classification and standards are established by industry organizations, such as the International Organization for Standardization (ISO) and various regulatory bodies. Here's an explanation of the importance of adhering to industry standards:

**Key Points:**

1. **ISO Standards:**

    The ISO has developed a series of standards, particularly ISO 14644-1, which provides guidelines for cleanroom classification based on airborne particle counts. Adhering to these standards ensures a consistent and widely accepted approach to cleanroom classification.

2. **Regulatory Compliance:**

    Different industries, such as pharmaceuticals, biotechnology, semiconductor manufacturing, and healthcare, may have specific regulations governing cleanroom practices. Adhering to industry standards ensures compliance with these regulations, avoiding potential legal and regulatory issues.

3. **Process Integrity:**

    Industry standards are designed to meet the specific cleanliness requirements of various processes. Adhering to these standards ensures that the cleanroom environment aligns with the needs of the processes conducted within, preventing contamination and maintaining process integrity.

4. **Quality Control:**

    Adherence to industry standards is crucial for maintaining product quality. Cleanroom processes are often sensitive to contaminants, and strict standards help prevent the introduction of particles that could compromise the quality of products, especially in industries like microelectronics and pharmaceuticals.

5. **Cross-Industry Consistency:**

   Industry standards provide a common language and framework for cleanroom classification. This consistency is essential for communication and collaboration between different organizations and industries. It allows for a standardized approach to cleanroom design, operation, and maintenance.

6. **Personnel Training:**

   Understanding and adhering to industry standards is part of personnel training. Cleanroom operators and staff must be aware of the specific requirements outlined in standards to ensure proper gowning, behavior, and adherence to cleanliness protocols.

7. **Facility Design and Operation:**

   Industry standards guide the design and operation of cleanroom facilities. This includes considerations such as air filtration systems, material selection, and monitoring protocols. Adhering to these standards helps create a controlled environment conducive to the intended processes.

8. **Continuous Improvement:**

   - Industry standards evolve over time to incorporate advancements in technology and best practices. Adhering to these standards ensures that cleanroom facilities stay current and can benefit from ongoing improvements in contamination control.

In summary, adhering to industry standards in cleanroom operations is vital for maintaining a controlled environment, meeting regulatory requirements, ensuring process integrity, and upholding product quality. It forms the foundation for effective and standardized cleanroom practices across diverse industries.

# Chapter 9. Cleanroom Access Control Orientation

## 9.1 Restricted access policies

Restricted access policies play a crucial role in maintaining the integrity of a cleanroom environment by controlling and monitoring who enters and exits the controlled space. Here's an explanation of the key elements related to restricted access policies under the Cleanroom Access Control Orientation:

**Key Points:**

1. **Personnel Authorization:**

   Restricted access policies define who is authorized to enter the cleanroom. Only personnel with a legitimate reason, such as those involved in specific processes or maintenance activities, are granted access. This authorization is typically determined based on job roles and responsibilities.

2. **Access Levels:**

   Access to different areas within the cleanroom may be restricted based on the level of cleanliness required. For example, more stringent access controls might be in place for critical manufacturing or testing areas compared to less critical support areas. Access levels are determined by the cleanroom classification and the sensitivity of ongoing processes.

3. **Identification and Authentication:**

   Personnel entering the cleanroom are required to undergo identification and authentication processes. This often involves using electronic key cards, biometric identification, or other secure methods to verify the identity of individuals. The goal is to ensure that only authorized personnel gain entry.

4. **Training and Certification:**

   Access to the cleanroom is typically granted only to personnel who have received proper training on cleanroom protocols and gowning procedures. Certification may be required to confirm that individuals understand and can adhere to the specific requirements of the cleanroom environment.

5. **Visitor Protocols:**

   Visitors, including contractors and suppliers, are subject to specific access protocols. They may need to be accompanied by authorized personnel, undergo orientation, and adhere to temporary access restrictions. Visitor access is closely monitored to prevent unauthorized entry.

6. **Monitoring and Surveillance:**

   Access control systems often include monitoring and surveillance features. This may involve security cameras, access logs, and real-time tracking of personnel movements within the cleanroom. Monitoring helps ensure compliance with access policies and provides a record of who entered the cleanroom and when.

7. **Emergency Access Procedures:**

   While access is restricted under normal circumstances, there should be clear procedures for emergency situations. Restricted access policies should include protocols for allowing emergency response teams or authorized personnel to quickly access the cleanroom in case of critical incidents.

8. **Periodic Reviews and Audits:**

   Restricted access policies should undergo periodic reviews and audits to ensure their effectiveness. This includes evaluating the list of authorized personnel, reviewing access logs, and updating access levels based on changes in processes or personnel responsibilities.

9. **Documentation and Communication:**

   Clear documentation of access control policies and effective communication to all personnel are essential. This ensures that everyone is aware of the access requirements, understands the consequences of unauthorized access, and can contribute to maintaining the integrity of the cleanroom environment.

10. **Enforcement and Consequences:**

    Strict enforcement of access control policies is critical. Consequences for violating these policies should be clearly communicated, and disciplinary measures may be implemented to deter unauthorized access.

In summary, restricted access policies are fundamental to maintaining the cleanliness and controlled environment of a cleanroom. These policies are designed to prevent contamination, protect ongoing processes, and ensure the overall effectiveness of the cleanroom facility.

## 9.2 Authentication procedures

Authentication procedures are a vital component of cleanroom access control, ensuring that only authorized individuals gain entry to the controlled environment. These procedures involve verifying the identity of personnel using secure and reliable methods. Here's an explanation of the key aspects of authentication procedures:

**Key Points:**

1. **Identification Methods:**

    Cleanroom access control systems employ various identification methods to authenticate individuals. Common methods include:

    - Electronic Key Cards: Personnel use electronically encoded key cards to gain access. Each card is unique to the individual and can be deactivated if lost or stolen.

    - Biometric Identification: Biometric data, such as fingerprints, retina scans, or facial recognition, may be used for precise and secure identification.

    - Personal Identification Numbers (PINs): Individuals may be required to enter a confidential PIN along with another identification method for dual authentication.

2. **Biometric Identification:**

    Biometric authentication relies on unique physical or behavioral characteristics of individuals. This could include fingerprint scans, iris or retina scans, facial recognition, or voice recognition. Biometric data is highly secure and difficult to replicate.

3. **Key Card Access:**

    Electronic key cards are commonly used for cleanroom access. These cards are encoded with specific information tied to the individual's identity and access level. Card readers at entry points authenticate the card and grant or deny access accordingly.

4. **PIN-Based Authentication:**

In addition to physical access cards, personnel may be required to enter a confidential Personal Identification Number (PIN) for further authentication. This adds an extra layer of security, especially when combined with other identification methods.

5. **Multi-Factor Authentication (MFA):**

Multi-factor authentication involves using two or more methods of authentication. For example, a combination of a key card and a fingerprint scan enhances security by requiring multiple forms of identification.

6. **Access Control Software:**

Authentication procedures are often managed through access control software. This software stores and processes information related to personnel identification, access levels, and entry logs. It allows administrators to configure and monitor access policies.

7. **Secure Communication Protocols:**

Communication between access control devices and the central system is secured using encryption protocols. This ensures that the authentication data transmitted is protected from unauthorized access or tampering.

8. **Integration with Personnel Databases:**

Cleanroom access control systems are often integrated with personnel databases. This integration ensures that access permissions are based on up-to-date information, including changes in personnel roles or responsibilities.

9. **Real-Time Authentication:**

Authentication processes happen in real-time to ensure immediate access decisions. This is crucial for maintaining a dynamic and secure cleanroom environment, especially in facilities with varying access levels.

10. **Lost Card or Credential Protocols:**

Procedures are established for reporting and addressing lost or stolen access cards or credentials promptly. Lost cards can be deactivated, preventing unauthorized access.

Authentication procedures are designed to provide a seamless yet secure process for individuals entering the cleanroom. By combining advanced identification methods and access control technologies, cleanroom facilities can maintain a high level of security while allowing authorized personnel to perform their duties efficiently.

# Chapter 10. Environmental Monitoring Orientation

## 10.1 Monitoring equipment usage

Environmental monitoring in a cleanroom is essential to ensure that the controlled environment meets specified cleanliness and operational standards. Monitoring equipment plays a crucial role in assessing and maintaining the conditions within the cleanroom. Here's an explanation of monitoring equipment usage:

**Key Points:**

1. **Particle Counters:**

   Particle counters are fundamental tools used to monitor airborne particulate contamination. These devices sample the air in the cleanroom and measure the concentration of particles of different sizes. Regular usage helps track cleanliness levels and identify trends that may require corrective actions.

2. **Airborne Particle Counting Systems:**

   These systems consist of particle counters strategically located throughout the cleanroom. They continuously monitor particle levels, providing real-time data. The information is often logged and can be used for trend analysis and to trigger alarms if particle levels deviate from specified limits.

3. **Microbial Monitoring Systems:**

   In cleanrooms where microbial contamination is a concern, microbial monitoring systems are employed. These systems sample air or surface swabs to detect the presence of microorganisms. Regular monitoring ensures compliance with microbial cleanliness standards.

4. **Temperature and Humidity Sensors:**

   Cleanrooms often have temperature and humidity sensors to monitor and control these environmental parameters. Continuous monitoring ensures that the cleanroom operates within the specified range, which is critical for certain processes and equipment.

5. **Differential Pressure Gauges:**

   Cleanrooms are designed with controlled pressure differentials to prevent the ingress of contaminants. Differential pressure gauges are used to measure and maintain these pressure differentials between adjoining areas. Monitoring ensures that the cleanroom maintains the required pressure differentials.

6. **Gas Analyzers:**

    In environments where specific gases need to be controlled, gas analyzers are used. These analyzers monitor the concentration of gases and ensure that they remain within acceptable limits.

7. **Liquid Particle Counters:**

    For cleanrooms involved in liquid-based processes, liquid particle counters are employed to monitor the cleanliness of liquids. This is crucial in applications where even small particles can impact product quality.

8. **Usage Calibration and Maintenance:**

    Regular calibration and maintenance of monitoring equipment are essential to ensure accurate and reliable measurements. Calibration verifies that the equipment is providing precise data, while maintenance prevents malfunctions that could compromise monitoring accuracy.

9. **Real-Time Monitoring Systems:**

    Advanced cleanrooms may use real-time monitoring systems that provide continuous data streams. These systems allow for immediate response to deviations from set parameters and enhance the ability to maintain a stable and controlled environment.

10. **Alarm Systems:**

    Monitoring equipment is often integrated with alarm systems. If any parameter deviates beyond acceptable limits, the alarm system triggers alerts to notify personnel. This allows for prompt corrective actions to maintain the cleanroom's integrity.

11. **Data Logging and Trend Analysis:**

    Monitoring equipment often includes data logging capabilities. This historical data can be analyzed to identify trends, assess the effectiveness of control measures, and make informed decisions about improvements or adjustments to the cleanroom environment.

Usage of monitoring equipment is critical for cleanroom operations, providing the necessary data to maintain a controlled environment conducive to sensitive manufacturing or research processes. Regular calibration, maintenance, and analysis of monitoring data contribute to the overall effectiveness of the cleanroom's environmental control.

## 10.2 Reporting deviations

In a cleanroom environment, maintaining stringent control over environmental conditions is crucial. Deviations from specified cleanliness levels or other environmental parameters may occur, and it is essential to promptly report such deviations to ensure corrective actions are taken. Here's an explanation of reporting deviations in the context of environmental monitoring:

**Key Points:**

1. **Immediate Reporting:**

   Personnel responsible for environmental monitoring should report any deviations from established standards immediately. This includes deviations in particle counts, microbial levels, temperature, humidity, pressure differentials, or any other monitored parameter.

2. **Identification of Deviations:**

   Regular monitoring activities involve the use of specialized equipment to measure and assess various environmental factors. When monitoring results indicate a deviation from the defined limits, personnel must be diligent in identifying and documenting these deviations.

3. **Use of Monitoring Records:**

   Monitoring records, which include data from particle counters, microbial monitoring systems, and other environmental sensors, serve as the basis for identifying deviations. Comparing current data with historical records helps recognize any abnormal trends or sudden spikes in contamination levels.

4. **Communication Protocols:**

   Clear communication protocols should be established to ensure that deviations are reported to relevant personnel promptly. This may involve notifying cleanroom supervisors, quality assurance teams, or designated individuals responsible for initiating corrective actions.

5.  **Documentation:**

- Deviations and the corresponding corrective actions should be thoroughly documented. This documentation serves as a record for regulatory compliance, quality assurance, and continuous improvement efforts. It includes details such as the nature of the deviation, time of occurrence, and the steps taken to address it.

6.  **Root Cause Analysis:**

Upon identifying a deviation, it is essential to conduct a root cause analysis to determine why the deviation occurred. This analysis helps in implementing corrective actions that address the underlying issues rather than just treating the symptoms.

7.  **Corrective Actions:**

Once a deviation is reported and analyzed, corrective actions should be implemented promptly. These actions may include adjusting equipment settings, recalibrating monitoring instruments, addressing procedural issues, or making physical changes to the cleanroom environment.

8.  **Preventive Measures:**

Reporting deviations is not only about addressing immediate concerns but also about implementing preventive measures. Identifying the root causes and implementing preventive actions help reduce the likelihood of similar deviations in the future.

9.  **Continuous Improvement:**

Reporting deviations is an integral part of the continuous improvement process. By learning from deviations and implementing improvements, cleanroom facilities can enhance their environmental control measures and maintain optimal conditions over the long term.

10. **Regulatory Compliance:**

In regulated industries, adherence to reporting procedures for deviations is critical for compliance with standards and regulations. Documentation of deviations and corrective actions is often subject to audit, and maintaining accurate records is essential.

Reporting deviations is a proactive and integral aspect of maintaining a controlled cleanroom environment. It ensures that any anomalies are addressed promptly, preventing potential negative impacts on product quality, research outcomes, or other critical processes within the cleanroom.

# Chapter 11. Documentation and Record Keeping Orientation

## 11.1 Maintaining accurate records

Maintaining accurate records is a fundamental aspect of cleanroom operations, contributing to regulatory compliance, quality assurance, and continuous improvement efforts. Here's an explanation of the importance and practices associated with maintaining accurate records in a cleanroom setting:

**Key Points:**

1. **Regulatory Compliance:**

   Accurate documentation is essential for complying with regulatory requirements in various industries, including pharmaceuticals, biotechnology, electronics manufacturing, and more. Regulatory bodies often mandate the recording of processes, procedures, and environmental conditions within cleanrooms.

2. **Quality Assurance:**

   Records serve as a critical component of quality assurance systems. They provide a detailed account of activities, processes, and environmental conditions, facilitating the tracking of product quality, process validation, and adherence to quality standards.

3. **Process Traceability:**

   Accurate records enable traceability of processes, ensuring that each step of a production or research process can be retraced. This is crucial for investigating issues, identifying root causes of deviations, and implementing corrective and preventive actions.

4. **Monitoring Environmental Conditions:**

   Cleanrooms require continuous monitoring of environmental parameters such as particle counts, temperature, humidity, and pressure differentials. Accurate recording of these conditions over time allows for trend analysis, early detection of deviations, and proactive maintenance.

5. **Equipment Calibration and Maintenance:**

   Records should be maintained for the calibration and maintenance of equipment used within the cleanroom. This includes tools, instruments, and systems critical for maintaining the controlled environment. Regular calibration ensures the accuracy and reliability of measurements.

6. **SOPs and Work Instructions:**

Standard Operating Procedures (SOPs) and work instructions are foundational documents in cleanroom environments. Accurate records of these documents, including version control and revisions, ensure that personnel follow approved and up-to-date procedures.

7. **Training and Certification Records:**

Personnel working in cleanrooms undergo training and certification. Accurate records of training sessions, certifications, and competency assessments are essential for ensuring that individuals are qualified and capable of performing their assigned tasks.

8. **Batch Records (Manufacturing):**

In manufacturing processes, especially in industries like pharmaceuticals, maintaining batch records is crucial. Batch records detail the steps taken during production, including raw material specifications, processing conditions, and quality control measures.

9. **Change Control Documentation:**

Changes to processes, equipment, or procedures are often part of continuous improvement initiatives. Maintaining accurate records of change control documentation ensures transparency, accountability, and a systematic approach to managing changes.

10. **Audit Trail:**

Accurate records create an audit trail, which is vital during regulatory inspections or internal audits. An audit trail provides a chronological sequence of events and actions, demonstrating compliance with established procedures and regulations.

11. **Data Integrity:**

Maintaining accurate records contributes to data integrity. This involves ensuring that data is complete, consistent, and accurate throughout its lifecycle. Data integrity is critical for making informed decisions based on reliable information.

12. **Continuous Improvement:**

Records serve as a foundation for continuous improvement initiatives. Analyzing historical data allows organizations to identify areas for improvement, implement corrective actions, and enhance overall cleanroom performance.

13. **Electronic Record-Keeping Systems:**

   With technological advancements, many cleanrooms use electronic record-keeping systems. These systems provide secure storage, easy retrieval, and efficient management of records, streamlining documentation processes.

In summary, maintaining accurate records is a cornerstone of effective cleanroom management. It supports compliance, quality assurance, and the pursuit of operational excellence by providing a documented history of processes, environmental conditions, and critical activities within the cleanroom.

## 11.2 Documenting processes and procedures

Documenting processes and procedures is a crucial aspect of cleanroom operations. This involves creating detailed and standardized documents that outline the steps, instructions, and guidelines for various activities within the cleanroom environment. Here's an explanation of the importance and practices associated with documenting processes and procedures in a cleanroom setting:

**Key Points:**

1. **Standardization:**

   Documenting processes and procedures ensures standardization in the execution of tasks within the cleanroom. Standardized procedures help maintain consistency, reduce variability, and enhance the overall quality of work.

2. **Compliance:**

   Cleanrooms operate within regulated industries, and adherence to specific processes and procedures is often mandated by regulatory authorities. Documentation provides evidence of compliance with regulatory requirements, facilitating audits and inspections.

3. **Training and Onboarding:**

   Detailed documentation serves as a valuable resource for training new personnel and onboarding them into the cleanroom environment. Standard operating procedures (SOPs) and work instructions provide step-by-step guidance, ensuring that employees understand and follow established protocols.

4. **Process Transparency:**

   Documenting processes provides transparency into the workflow and sequence of activities. This transparency is essential for understanding how different tasks are interconnected and how each step contributes to the overall operation of the cleanroom.

5. **Risk Mitigation:**

   Clearly documented processes help identify potential risks and hazards associated with specific activities. Risk assessments can be conducted, and mitigation strategies can be implemented to enhance safety and prevent incidents.

6. **Quality Control:**

   Documenting processes facilitates quality control by specifying critical parameters, acceptance criteria, and quality checkpoints. This ensures that each step of a process is monitored, and the final output meets predetermined quality standards.

7. **Efficiency and Productivity:**

   Well-documented processes contribute to increased efficiency and productivity. Employees can follow established procedures, reducing the likelihood of errors, rework, and delays. This, in turn, enhances the overall output and performance of the cleanroom.

8. **Change Control:**

   Processes and procedures may undergo changes over time due to improvements, updates, or regulatory requirements. Documenting changes through a formal change control process ensures that all stakeholders are informed, and the revised documents are accessible to relevant personnel.

9. **Validation and Qualification:**

   In industries such as pharmaceuticals, validation and qualification are critical. Documenting processes is essential for demonstrating that processes have been validated and equipment has been qualified to operate within specified parameters.

10. **Traceability:**

Documenting processes provides traceability, allowing organizations to trace the history and evolution of procedures. This is valuable for troubleshooting, identifying root causes of issues, and implementing corrective actions.

11. **Cross-Functional Collaboration:**

Cleanroom operations often involve collaboration among different departments and teams. Documented processes serve as a common reference point, facilitating effective communication and collaboration among cross-functional teams.

12. **Continuous Improvement:**

Documented processes support continuous improvement initiatives by providing a baseline for evaluating current practices. Regular reviews and updates to documentation enable organizations to adapt to changing requirements and implement best practices.

13. **Electronic Documentation Systems:**

Many cleanrooms utilize electronic documentation systems for efficient management and storage of process documents. Electronic systems provide version control, accessibility, and security for maintaining and updating documentation.

In summary, documenting processes and procedures in a cleanroom environment is essential for ensuring consistency, compliance, and operational excellence. It plays a pivotal role in training, risk management, quality control, and overall efficiency within the cleanroom setting.

# Chapter 12 Managing Emergencies

## 12.1 HF Spill on somebody's skin

Handling hydrofluoric acid (HF) spills on the skin requires immediate and specific actions due to the severe toxicity of HF. HF can cause serious chemical burns and systemic effects. If someone is exposed to HF, here are the recommended steps:

1. **Emergency Response:**

   Call for emergency medical assistance immediately. HF exposure requires prompt medical attention.

2. **Evacuation:**

   Remove the affected person from the area where the exposure occurred to prevent further contact with HF.

3. **Protect Yourself:**

   Put on appropriate personal protective equipment (PPE), such as gloves and safety goggles, before attempting to help the affected person.

4. **First Aid Measures:**

   Rinse the affected area with water immediately. Use a safety shower or an eyewash station if available. Continue rinsing the affected area for at least 15 minutes. Prompt and prolonged rinsing is crucial to remove HF from the skin.

5. **Remove Contaminated Clothing:**

   If clothing is contaminated with HF, remove it while rinsing to prevent further exposure.

6. **Seek Medical Attention:**

   Transport the affected person to the nearest medical facility or call for emergency medical services. HF exposure requires professional medical evaluation and treatment.

7. **Provide Information to Medical Professionals:**

   Provide information about the HF exposure to medical professionals. Include details such as the concentration of HF, the duration of exposure, and any first aid measures taken.

8. **Do Not Use Neutralizing Agents:**

   Avoid using neutralizing agents (such as calcium gluconate) without professional medical guidance. HF requires specialized medical treatment, and neutralizing agents may not be suitable.

Remember, HF exposure can lead to serious health effects, including tissue damage, systemic toxicity, and potentially fatal outcomes. It is crucial to seek immediate medical attention, and only trained medical professionals should handle HF exposures.

Note: The information provided here is general in nature and should not be considered a substitute for professional medical advice. Always follow specific safety procedures and consult with medical professionals in the event of a chemical exposure.

## 12.2 HF Spill on Cleanroom Floor

Certainly, when dealing with an HF spill on the floor in a cleanroom, a spill response kit designed for handling HF incidents is crucial. Such kits typically include materials and instructions for safe cleanup. Here's a general guideline for responding to an HF spill using a spill response kit:

1. **Emergency Response:**

   Alert personnel in the area and evacuate if necessary. Follow the cleanroom's emergency procedures.

2. **Personal Protective Equipment (PPE):**

   Put on the appropriate PPE provided in the spill response kit. This may include acid-resistant gloves, goggles or face shield, and a chemical-resistant apron.

3. **Isolate the Area:**

   Clearly mark and isolate the contaminated area using barriers or signs to prevent accidental exposure.

4. **Ventilation:**

   Ensure the cleanroom's ventilation is functioning properly. If the spill is in an area with local exhaust ventilation, activate it to help disperse any vapors.

5. **Spill Response Kit Usage:**

Follow the instructions provided in the spill response kit. Kits may contain absorbent materials specifically designed for HF spills, neutralizing agents, and tools for cleanup.

6. **Containment and Cleanup:**

Use the absorbent materials from the spill response kit to contain and absorb the spilled HF. Carefully follow the kit's instructions for cleanup.

7. **Neutralization (if applicable):**

If the spill response kit includes a neutralizing agent, apply it as directed. Follow the kit's guidelines for neutralization, considering the type and concentration of the HF.

8. **Avoid Creating Aerosols:**

Minimize the generation of aerosols during cleanup to prevent the spread of HF. Use controlled movements and techniques as recommended in the spill response kit.

9. **Dispose of Contaminated Materials:**

Dispose of all contaminated materials, including used absorbents and cleaning materials, following the cleanroom's waste disposal procedures and relevant regulations.

10. **Decontamination:**

Decontaminate the affected area thoroughly, following cleanroom protocols. This may involve multiple cleaning steps.

11. **Document the Incident:**

Record details of the spill, response actions, and observations. Documenting the incident is essential for review and improvement of cleanroom safety procedures.

Always consult the specific guidelines provided in the spill response kit and adhere to the cleanroom's safety protocols. If in doubt or if the spill is substantial, seek assistance from the cleanroom's safety officer or emergency response team.

## 12.3 Describe How to Use the HF Spill Kit

Using an HF (hydrofluoric acid) spill kit requires careful adherence to safety protocols. Here's a detailed description of how to use the HF spill kit:

1. **Assess the Situation:**

   Before approaching the spill, evaluate the size and severity of the HF spill. Ensure you have the appropriate personal protective equipment (PPE) on.

2. **Put on Personal Protective Equipment (PPE):**

   Wear acid-resistant gloves, goggles or a face shield, and a chemical-resistant apron before attempting to handle the spill. This step is crucial to protect yourself from HF exposure.

3. **Isolate the Area:**

   If possible, cordon off the affected area to prevent unauthorized personnel from entering. Post warning signs indicating a hazardous material spill.

4. **Retrieve the HF Spill Kit:**

   Locate the designated HF spill kit in the cleanroom. Kits are usually strategically placed for easy access in case of an emergency.

5. **Verify Kit Contents:**

   Check the contents of the HF spill kit to ensure it contains the necessary items:

   - Absorbent materials designed for HF spills (e.g., calcium carbonate or other HF-specific neutralizers).
   - Personal protective equipment (PPE): acid-resistant gloves, goggles or face shield, chemical-resistant apron.
   - Tools for containment and cleanup (e.g., scoops, shovels).
   - Instructions and guidelines specific to HF spill response.

6. **Read Instructions:**

   Carefully read the step-by-step instructions provided in the spill kit. Follow the guidelines precisely to ensure a safe and effective response.

7. **Neutralize and Absorb the Spill:**

   Use the provided absorbent materials to carefully cover the spilled HF. Neutralize the acid by applying the absorbent material and allowing it to react with the HF.

8. **Use Tools for Containment:**

> Utilize the tools provided in the kit, such as scoops or shovels, to carefully contain and collect the neutralized spill. Avoid generating aerosols or splashes.

9. **Dispose of Waste:**

> Place the absorbed material into appropriate waste containers designated for hazardous materials. Follow proper waste disposal procedures in accordance with cleanroom protocols and regulations.

10. **Decontaminate Tools:**

> If reusable tools were used, decontaminate them following the provided guidelines before returning them to the kit.

11. **Document the Response:**

> Record all actions taken during the HF spill response. This documentation is essential for post-incident analysis and continuous improvement of safety protocols.

Using the HF spill kit requires precision and adherence to safety guidelines to minimize the risks associated with handling this hazardous material in a cleanroom environment.

## 12.4 Piranha Spill on somebody's Skin

**Responding to a Piranha solution spill on someone's skin requires immediate and careful action due to the highly corrosive nature of Piranha. Here is the correct response:**

1. **Priority:**

> Prioritize the safety of the affected individual. Act swiftly, as exposure to Piranha solution can cause severe chemical burns.

2. **Emergency Eyewash and Shower:**

> If available, move the affected person to the nearest emergency eyewash and shower station. Rinse the affected area thoroughly with copious amounts of water. This helps dilute and remove the corrosive chemical.

3. **Remove Contaminated Clothing:**

   While rinsing, remove any clothing or accessories that may have come into contact with the Piranha solution. Cut the clothing if necessary to avoid further skin exposure.

4. **Continue Rinsing:**

   Continue rinsing the affected area with water for at least 15 minutes or as per safety guidelines. Ensure that the water reaches all exposed skin surfaces.

5. **Seek Medical Attention:**

   Immediately seek medical attention or emergency medical services. Even if the individual does not show immediate signs of injury, Piranha solution can cause delayed and severe tissue damage.

6. **Emergency Response Team:**

   Activate the hazardous material response team or emergency response team as per the cleanroom's protocols. They are trained to handle chemical spills and provide additional assistance.

7. **First Aid:**

   While waiting for medical professionals, administer basic first aid if you are trained to do so. This may include covering the affected area with a clean and non-stick bandage.

8. **Provide Information:**

   Share detailed information about the spilled substance (Piranha solution) with medical professionals to ensure appropriate and prompt treatment.

9. **Document the Incident:**

   Document the incident, including the details of the spill and the response actions taken. This documentation is crucial for analysis and improvement of safety protocols.

10. **Review and Improve Procedures:**

   After the incident, conduct a thorough review of the response procedures. Identify areas for improvement and implement changes to prevent similar incidents in the future.

Remember, safety is paramount, and immediate action is crucial in the event of a Piranha solution spill on the skin. Always follow established safety protocols and seek professional medical assistance promptly.

## 12.5 Piranha Spill on Cleanroom Floor

Responding to a Piranha solution spill on a cleanroom floor is a critical task to ensure the safety of personnel and prevent contamination. Here are the steps to address a Piranha spill on a cleanroom floor:

1. **Evacuate the Area:**

   Clear the immediate area of personnel to prevent exposure. Use alarms or communication systems to notify everyone in the vicinity of the spill to evacuate promptly.

2. **Emergency Response Team Activation:**

   Activate the hazardous material response team or emergency response team as per the cleanroom's protocols. Trained professionals equipped with appropriate personal protective equipment (PPE) should handle the cleanup.

3. **Isolate the Spill Area:**

   Use barricades or signs to isolate the spill area and prevent unauthorized access. This helps contain the potential spread of the hazardous substance.

4. **Personal Protective Equipment (PPE):**

   Ensure that responders are equipped with the necessary PPE, including chemical-resistant gloves, goggles or face shields, and protective clothing to prevent skin contact.

5. **Neutralization:**

   If there are designated neutralizing agents for Piranha spills in the cleanroom, apply them cautiously and following prescribed procedures. Neutralizing agents should be specifically recommended for Piranha solutions.

6. **Absorbent Materials:**

   Use absorbent materials, such as spill pillows, to contain and absorb the spilled Piranha solution. Carefully place the absorbent materials over the spill, working from the outer edges toward the center.

7.  **Chemical Spill Kit:**

> Utilize a chemical spill response kit designed for hazardous materials like Piranha solutions. Follow the instructions in the kit for safe and effective cleanup.

8.  **Proper Disposal:**

> Collect the absorbed Piranha solution along with any contaminated materials in appropriate containers for hazardous waste disposal. Follow established procedures for hazardous waste management.

9.  **Decontamination:**

> Decontaminate the affected area thoroughly using recommended cleaning agents. Ensure that residues are completely removed to prevent future exposure risks.

10. **Review and Document:**

> After the cleanup, conduct a thorough review of the incident. Document the spill response, cleanup procedures, and any lessons learned. This information is valuable for improving safety protocols.

11. **Training and Preparedness:**

> Ensure that cleanroom personnel are trained on spill response procedures and understand the importance of immediate reporting of spills. Regular drills and training sessions contribute to preparedness.

Remember, Piranha solutions are highly corrosive and pose significant risks. It is essential to involve trained personnel with appropriate safety equipment and follow established cleanroom protocols for hazardous material spills.

## 12.6 Can Cleanroom team use HF Spill kit for Piranha?

HF (hydrofluoric acid) and Piranha solution spills require different response strategies due to their distinct properties. While there may be similarities in the initial steps of responding to acid spills, the neutralization and cleanup procedures differ.

For HF spills, a specific HF spill response kit is recommended. HF is a highly corrosive acid, and traditional spill kits designed for general chemical spills may not be effective. HF spill kits typically include calcium gluconate gel, which is used for skin exposure to neutralize the acid.

On the other hand, Piranha solution, which is a mixture of concentrated sulfuric acid and hydrogen peroxide, poses additional challenges. Piranha solution is an oxidizer and can react violently with organic materials. The response to Piranha spills involves neutralization and careful cleanup to prevent further reactions.

Therefore, it's crucial to use a spill response kit that is specifically designed for Piranha solutions. This kit should include materials and agents suitable for handling the specific hazards associated with Piranha spills.

In summary:

- **For HF spills:** Use an HF spill response kit containing calcium gluconate gel and other materials designed for HF neutralization.

- **For Piranha solution spills:** Use a spill response kit specifically designed for handling the hazards of Piranha solutions, which may include neutralizing agents suitable for sulfuric acid and hydrogen peroxide.

Always follow the instructions provided in the kits and adhere to the safety protocols established for your cleanroom or laboratory. If in doubt, consult with safety professionals or specialists familiar with the handling of hazardous materials in your specific environment.

## 12.7 Gas Leakage

In the event of a gas leakage in a cleanroom, the cleanroom manager/engineer/coordinator should follow a set of predefined emergency response procedures to ensure the safety of personnel, protect the cleanroom environment, and mitigate potential hazards. Here is a general guideline for the correct response:

1. **Initiate Emergency Procedures:**

   - Immediately activate the cleanroom's emergency response plan.

   - Use the facility's alarm system to alert personnel about the gas leakage. This may involve audible alarms, visual signals, or automated messages.

2. **Evacuation:**

   - In case of a significant gas leak, initiate evacuation procedures. Clear the cleanroom of all personnel to a designated safe assembly area.

   - Ensure that personnel are aware of the evacuation routes and assembly points.

3. **Isolate the Area:**

   If possible and without compromising safety, isolate the affected area by closing doors or using other containment measures to prevent the spread of the gas.

4. **Notify Emergency Services:**

   Contact emergency services, such as the fire department or relevant authorities, to report the gas leakage. Provide them with accurate and detailed information about the type of gas, its potential risks, and the measures taken to contain the situation.

5. **Communication:**

   Maintain communication with all personnel to keep them informed about the situation and provide guidance on necessary actions.

   Establish a communication point for emergency responders to coordinate efforts.

6. **Safety Gear and Personal Protection:**

   Ensure that personnel involved in the response, including the cleanroom manager/engineer/coordinator and emergency responders, wear appropriate personal protective equipment (PPE) based on the nature of the gas.

7. **Assessment and Monitoring:**

   - Use gas detection equipment to assess the concentration of the leaked gas and monitor the cleanroom environment.

   - Regularly update emergency responders and personnel on the progress of the response and any changes in the situation.

8. **Cleanroom Integrity Check:**

   Once the gas leakage is under control, conduct a thorough check of the cleanroom's integrity to ensure that it is safe for re-entry.

9. **Investigation and Documentation:**

   Initiate an investigation into the cause of the gas leakage and document the incident for analysis and improvement of future safety protocols.

It's important to note that the specific response may vary depending on the type of gas, cleanroom specifications, and established safety protocols. Cleanroom managers should ensure that personnel are well-trained in emergency procedures and regularly conduct drills to reinforce preparedness.

## 12.8 Sudden Power Shortage

In the event of a sudden power shortage or blackout in a cleanroom, it's crucial to follow established procedures to ensure the safety of personnel, protect sensitive equipment, and maintain the integrity of the cleanroom environment. Here is a general guideline for the correct response:

1. **Initiate Emergency Procedures:**

   - Immediately activate the cleanroom's emergency response plan specific to power shortages.

   - Use the facility's alarm system to notify personnel about the power outage.

2. **Assess the Situation:**

   - Quickly assess the impact of the power shortage on critical operations, equipment, and safety systems.

   - Identify any immediate risks or hazards associated with the loss of power.

3. **Emergency Lighting:**

   Ensure that emergency lighting systems are activated to provide sufficient illumination for safe evacuation and movement within the cleanroom.

4. **Communicate with Personnel:**

   - Maintain communication with all personnel to keep them informed about the situation and provide guidance on necessary actions.

   - Instruct personnel to remain calm and follow established evacuation procedures if required.

5. **Backup Power Systems:**

   - Check if the cleanroom is equipped with backup power systems, such as uninterruptible power supply (UPS) units or emergency generators.

   - If available, activate backup power systems to support critical operations and maintain essential services.

6. **Shutdown Procedures:**

   - Initiate shutdown procedures for sensitive equipment and processes to prevent damage when power is restored.

   - Follow established protocols for safely shutting down and restarting equipment.

7. **Emergency Exit Procedures:**

   - Remind personnel of emergency exit procedures and evacuation routes.

   - If necessary, guide personnel to exit the cleanroom safely using emergency exits and assembly points.

8. **Coordinate with Utility Services:**

   Contact the utility provider or facility management to report the power outage and receive updates on the estimated time for power restoration.

9. **Recovery and Restart:**

   - Once power is restored, follow documented procedures for systematically restarting cleanroom equipment and systems.

   - Conduct checks to ensure that critical parameters are within acceptable limits before resuming normal operations.

10. **Investigation and Documentation:**

    Investigate the cause of the power shortage and document the incident for analysis and improvement of future emergency response plans.

It's essential for cleanroom personnel to be well-trained in responding to power outages, and regular drills should be conducted to enhance preparedness. Additionally, cleanroom managers should periodically review and update emergency response plans to address any changes in equipment or facility configurations.

# Chapter 13: Cleanroom Operations Management

## 13.1 Cleanroom Daily Housekeeping Procedure

Daily cleanroom housekeeping is a crucial aspect of maintaining a controlled and contamination-free environment. It involves routine cleaning practices to eliminate particles, residues, and other contaminants that could compromise the integrity of the cleanroom. Here are key elements of daily cleanroom housekeeping:

1. **Surface Cleaning:**

   - Wipe down all surfaces, including walls, ceilings, and equipment, using lint-free and non-shedding cleanroom wipes or mop heads.

   - Pay special attention to critical surfaces and areas prone to particle accumulation.

2. **Floor Cleaning:**

   - Vacuum or mop the cleanroom floor using approved cleaning solutions. The cleaning method may depend on the cleanroom classification and the level of cleanliness required.

   - Ensure that the cleaning process minimizes the generation of airborne particles.

3. **Equipment Cleaning:**

   - Clean and disinfect equipment surfaces regularly, especially those in direct contact with products or processes.

   - Follow manufacturer-recommended cleaning procedures for specific equipment.

4. **Gowning Area Maintenance:**

   - Keep gowning areas clean and well-maintained to prevent the transfer of contaminants from personnel to the cleanroom environment.

   - Provide adequate storage for cleanroom garments and personal items.

5. **Air Shower and Entryway Maintenance:**

   - Regularly clean and disinfect air shower chambers and entryway mats to prevent the introduction of contaminants into the cleanroom.

   - Ensure that personnel follow proper gowning procedures before entering the cleanroom.

6. **Cleaning Agents and Disinfectants:**

   - Use cleanroom-compatible cleaning agents and disinfectants to minimize the risk of introducing chemical contaminants.

   - Ensure that cleaning agents do not leave residues that could impact the cleanliness of the environment.

7. **Controlled Waste Disposal:**

   - Dispose of waste, including used wipes, gloves, and other disposable items, in accordance with cleanroom waste disposal procedures.

   - Segregate and properly label waste containers to facilitate safe disposal.

8. **Regular Inspections:**

   - Conduct regular inspections to identify potential sources of contamination or cleanliness issues.

   - Address any deviations or non-compliance promptly to maintain the desired cleanliness level.

9. **Documentation:**

   - Maintain detailed records of daily cleanroom housekeeping activities, including cleaning schedules, cleaning agents used, and any deviations or issues identified.

   - Document any corrective actions taken to address cleanliness concerns.

10. **Training and Awareness:**

   - Provide ongoing training to cleanroom personnel on proper housekeeping procedures and the importance of maintaining a clean environment.

   - Foster a culture of cleanliness and awareness to ensure that all team members contribute to maintaining the desired cleanliness standards.

By consistently following these daily cleanroom housekeeping practices, organizations can uphold the required cleanliness levels, adhere to industry standards, and minimize the risk of contamination in critical environments.

## 13.2 Cleanroom Stocking Check

Checking the stocking of consumables and chemicals in a cleanroom involves regular assessments to ensure an adequate and organized inventory. The frequency of these checks may vary based on the usage patterns, criticality of the materials, and cleanroom requirements. Here's a general guideline:

**Regular Checkpoints:**

1. **Daily Checks:**

   - **Critical Consumables:** Items critical to daily operations, such as gloves, wipes, and swabs, should be checked daily to ensure an ample supply.

   - **Chemical Usage:** If certain chemicals are used daily, their stock levels should be reviewed regularly to avoid running out unexpectedly.

2. **Weekly Checks:**

   - **Medium-Usage Consumables:** Items with moderate usage, such as cleaning solutions and disposable garments, can be checked on a weekly basis.

   - **Regular Calibration Gases:** For cleanrooms with gas monitoring systems, weekly checks on the availability of calibration gases may be necessary.

3. **Monthly Checks:**

   - **Low-Usage Consumables:** Items with low consumption rates, such as less frequently used tools or specific cleaning agents, can be checked on a monthly basis.

   - **Bulk Chemicals:** Large quantities of chemicals stored in bulk containers may be assessed monthly to ensure an adequate supply.

**Procedure for Checking Stocking:**

1. **Create a Checklist:**

   Develop a checklist that includes all critical consumables, chemicals, and other items essential for cleanroom operations.

2. **Storage Area Inspection:**

   - Regularly inspect the storage areas to ensure proper organization and cleanliness.

   - Check for expired or outdated consumables and chemicals and remove them as needed.

3. **Usage Tracking:**

   Maintain a log or record of consumable usage to anticipate when replenishment will be necessary.

4. **Communication:**

   Establish effective communication channels between cleanroom staff and inventory personnel to relay information about low stock levels or urgent needs.

5. **Ordering Process:**

   Streamline the ordering process to ensure prompt replenishment. Consider setting up automatic reordering for items with predictable consumption patterns.

6. **Vendor Relationships:**

   Maintain good relationships with vendors to facilitate quick deliveries and minimize delays in restocking.

7. **Emergency Supplies:**

   Keep a reserve of emergency supplies to address unexpected spikes in demand or delays in restocking.

8. **Documentation:**

   Document each stock check, noting any deviations, and maintain records for compliance and auditing purposes.

9. **Continuous Improvement:**

   Periodically review the stocking procedures and make adjustments based on usage patterns, changes in operations, or feedback from cleanroom personnel.

By implementing a systematic approach to checking consumables and chemicals, organizations can ensure that cleanrooms are well-equipped, operational disruptions are minimized, and the risk of running out of critical supplies is mitigated. The frequency of checks can be adjusted based on the specific needs and dynamics of the cleanroom environment.

## 13.3 Cleanroom Inventory Control

Cleanroom inventory control is a critical aspect of maintaining the operational efficiency and compliance of a cleanroom environment. The process involves systematic management, monitoring, and documentation of all items within the cleanroom, including consumables, equipment, and materials. Here's an overview of the cleanroom inventory control procedure:

**Cleanroom Inventory Control Procedure:**

1. **Establish an Inventory Management System:**

   - Implement a robust inventory management system that is tailored to the specific needs of the cleanroom.

   - Utilize specialized inventory software or systems that allow for real-time tracking and reporting.

2. **Categorize Inventory Items:**

   - Classify inventory items into categories based on criticality, usage frequency, and specific cleanroom requirements.

   - Categories may include consumables, chemicals, tools, equipment, and spare parts.

3. **Define Stocking Levels:**

   - Establish minimum and maximum stocking levels for each category of inventory items.

   - Set reorder points to ensure timely replenishment and prevent stockouts.

4. **Barcoding or RFID Tagging:**

   - Implement barcoding or RFID tagging for easy identification and tracking of inventory items.

   - Use scanning devices to streamline data capture and update inventory records.

5. **Storage Area Organization:**

   - Organize storage areas within the cleanroom based on the categorization of items.

   - Clearly label storage locations and shelves to facilitate quick retrieval.

6. **Regular Stock Checks:**

   - Conduct regular physical stock checks to verify the accuracy of inventory records.

- Perform cycle counting or periodic audits to ensure consistency.

7. **Usage Tracking:**

   - Track the usage of consumables and materials in real-time.

   - Maintain usage logs to analyze patterns and forecast future needs.

8. **Documentation and Record Keeping:**

   - Document all inventory-related transactions, including receipts, issuances, and returns.

   - Keep accurate records for compliance, auditing, and traceability purposes.

9. **Supplier Relationship Management:**

   - Establish strong relationships with suppliers to ensure reliable and timely deliveries.

   - Monitor supplier performance and address any issues promptly.

10. **Emergency Preparedness:**

    - Develop contingency plans for unexpected spikes in demand or supply chain disruptions.

    - Maintain emergency stock levels for critical items.

11. **Training and Awareness:**

    - Train cleanroom personnel on the proper procedures for requesting, issuing, and returning inventory items.

    - Foster awareness about the importance of accurate inventory control.

12. **Continuous Improvement:**

    - Regularly review and refine the inventory control process based on feedback, changing requirements, and technological advancements.

    - Implement continuous improvement initiatives to enhance efficiency.

By following a comprehensive cleanroom inventory control procedure, organizations can ensure that the cleanroom remains well-stocked, operational disruptions are minimized, and compliance with regulatory standards is maintained. The use of advanced inventory management tools and technologies contributes to the accuracy and efficiency of the overall process.

## 13.4 How to Review a Cleanroom Construction Proposal?

Reviewing a cleanroom proposal involves a careful examination of various aspects to ensure that the proposed plan aligns with the specific needs and requirements of the cleanroom facility. Here is a step-by-step guide on how to review a cleanroom proposal:

**1. Understand Project Objectives:**

- **Define Cleanroom Purpose:**

    - Clearly understand the intended use and purpose of the cleanroom.

    - Verify that the proposal addresses specific industry standards and regulations applicable to the cleanroom's intended function.

**2. Evaluate Design and Layout:**

- **Cleanroom Classifications:**

    - Check if the proposed cleanroom classification aligns with the project requirements.

    - Ensure that the design meets the necessary cleanliness standards for the intended application.

- **Layout Efficiency:**

    - Evaluate the efficiency of the cleanroom layout.

    - Confirm that the proposed design optimizes workflow, minimizes contamination risks, and allows for easy access to equipment and workstations.

**3. Assess HVAC Systems:**

- **Airflow Control:**

    - Review the proposed HVAC (Heating, Ventilation, and Air Conditioning) system to ensure proper airflow control.

    - Confirm that the system can maintain the required cleanliness levels and control temperature and humidity.

- **Filtration Systems:**

    - Verify the specifications of air filtration systems.

    - Ensure that the proposed filtration systems are appropriate for the cleanroom classification and application.

## 4. Check Material and Construction:

- **Material Selection:**
    - Evaluate the materials proposed for cleanroom construction, including walls, ceilings, and flooring.
    - Confirm that selected materials meet cleanliness and durability requirements.
- **Construction Methods:**
    - Assess the proposed construction methods to ensure they comply with industry standards.
    - Confirm that construction practices minimize the risk of contamination during the building process.

## 5. Review Cleanroom Equipment:

- **Compatibility:**
    - Check the compatibility of proposed equipment with the cleanroom's intended use.
    - Ensure that equipment meets industry standards and is designed for cleanroom environments.
- **Maintenance Requirements:**
    - Evaluate the maintenance requirements of cleanroom equipment.
    - Confirm that the proposal includes plans for routine maintenance and calibration.

## 6. Consider Cleanroom Validation:

- **Validation Protocols:**
    - Verify that the proposal includes detailed validation protocols.
    - Ensure that the validation process aligns with industry standards and will be performed by qualified professionals.
- **Testing and Certification:**
    - Confirm that the proposal outlines testing and certification procedures for the cleanroom.
    - Ensure that certification is provided by accredited agencies.

## 7. Evaluate Compliance and Documentation:

- **Regulatory Compliance:**

    - Check if the proposal addresses regulatory compliance and industry standards.

    - Confirm that the cleanroom design and construction adhere to local, national, and international regulations.

- **Documentation Requirements:**

    - Ensure that the proposal includes comprehensive documentation requirements.

    - Confirm that documentation covers design specifications, construction plans, and validation reports.

## 8. Consider Safety Protocols:

- **Safety Features:**

    - Review safety features proposed for the cleanroom.

    - Confirm that safety protocols are in place for emergency situations, hazardous material handling, and personnel protection.

## 9. Evaluate Cost Estimates:

- **Budgetary Considerations:**

    - Review the proposed cost estimates for the entire cleanroom project.

    - Assess the cost breakdown, including construction, equipment, validation, and ongoing maintenance.

## 10. Seek Expert Input:

- **Engage Experts:**

    - If needed, consult with cleanroom design and construction experts for a thorough evaluation.

    - Seek input from professionals with experience in the specific industry and cleanroom requirements.

## Conclusion:

A comprehensive review of a cleanroom proposal involves a multidisciplinary approach, considering design, construction, equipment, validation, and compliance aspects. Engaging with experts and stakeholders, and ensuring alignment with industry standards, will contribute to the successful implementation of the cleanroom project.

## 13.5 Describe How to Make the Cleanroom Validation

Cleanroom validation is a critical process to ensure that a cleanroom facility meets the specified standards and performs as intended. The validation process typically involves a series of tests and evaluations to confirm that the cleanroom meets cleanliness, environmental control, and operational requirements. Here's a step-by-step guide on how to perform cleanroom validation:

**1. Establish Validation Protocols:**

- **Define Validation Criteria:**

    - Clearly define the criteria that the cleanroom must meet for validation.

    - Determine the required cleanliness level, temperature, humidity, and other environmental parameters.

- **Refer to Industry Standards:**

    - Consult relevant industry standards and regulations to ensure compliance.

    - Develop validation protocols based on recognized guidelines such as ISO 14644 for cleanroom classification.

**2. Pre-Validation Planning:**

- **Document Validation Plan:**

    - Develop a comprehensive validation plan outlining the scope, objectives, and methods.

    - Define the testing schedule and allocate resources for the validation process.

- **Coordinate with Stakeholders:**

    - Collaborate with relevant stakeholders, including cleanroom designers, contractors, and facility managers.

    - Ensure that everyone involved understands the validation process and their respective responsibilities.

### 3. Installation Qualification (IQ):

- **Verify Installation Compliance:**
    - Confirm that the cleanroom has been constructed according to the approved design specifications.
    - Check that all components, systems, and equipment are installed correctly.
- **Review Documentation:**
    - Examine documentation related to equipment manuals, drawings, and specifications.
    - Ensure that all installation records are complete and accurate.

### 4. Operational Qualification (OQ):

- **Functional Testing:**
    - Perform functional testing of cleanroom systems and equipment.
    - Verify that each component operates according to its design and functional requirements.
- **Environmental Control Checks:**
    - Validate environmental control parameters such as temperature, humidity, and airflow.
    - Confirm that the cleanroom maintains the specified cleanliness level.

### 5. Performance Qualification (PQ):

- **Simulate Real Operating Conditions:**
    - Introduce personnel and equipment into the cleanroom to simulate real operating conditions.
    - Monitor the cleanroom's performance during regular operations.
- **Evaluate Contingency Measures:**
    - Assess the effectiveness of contingency measures, such as emergency shutdown procedures.
    - Ensure that the cleanroom can recover from deviations or unexpected events.

### 6. Particle Count Testing:

- **Particle Count Measurements:**
    - Conduct particle count testing using a particle counter.
    - Sample air at various locations and classify cleanliness based on particle count limits.

## 7. Microbial Monitoring:

- **Microbial Air and Surface Sampling:**
    - Perform microbial monitoring through air and surface sampling.
    - Evaluate microbial contamination levels to ensure compliance with specified limits.

## 8. Airflow Visualization:

- **Smoke Studies:**
    - Conduct airflow visualization studies using smoke or other visualization agents.
    - Verify that airflows follow the intended paths and prevent cross-contamination.

## 9. Documentation and Reporting:

- **Record and Document Results:**
    - Document all testing and evaluation results, including any deviations or non-conformities.
    - Maintain detailed records for future reference and auditing.
- **Generate Validation Report:**
    - Compile all validation data and findings into a comprehensive validation report.
    - Clearly outline whether the cleanroom has passed validation and is ready for routine operations.

## 10. Review and Approval:

- **Internal Review:**
    - Conduct an internal review of the validation results with the validation team and relevant stakeholders.
    - Address any outstanding issues or concerns.

- **External Review (if applicable):**

    - Engage external validation experts or regulatory agencies for an independent review.

    - Seek necessary approvals or certifications for compliance.

## 11. Regular Monitoring and Maintenance:

- **Implement Regular Monitoring:**

    - Establish a routine monitoring program to ensure ongoing compliance with cleanliness and environmental requirements.

    - Schedule periodic re-validation as needed.

- **Address Changes or Upgrades:**

    - If there are changes or upgrades to the cleanroom, perform additional validation to ensure continued compliance.

## Conclusion:

Cleanroom validation is an integral part of the cleanroom lifecycle, ensuring that the facility consistently meets the required standards for cleanliness and environmental control. The process requires careful planning, adherence to protocols, and collaboration among stakeholders to achieve successful validation and maintain the cleanroom's operational integrity.

# Chapter 14 Cleanroom Management Skills

## 14.1 How to Make a Standard Operating Procedure?

Creating a Standard Operating Procedure (SOP) for equipment, tools, or devices in a cleanroom is crucial for ensuring consistent and safe operations. Below is a step-by-step guide on how to develop a Standard Operating Procedure for cleanroom equipment:

1. **Define the Purpose:**

   Clearly articulate the purpose of the SOP. What equipment or tool does it cover? What specific processes or tasks does it address? Understanding the scope is essential.

2. **Identify the Responsible Personnel:**

   Clearly define who is responsible for operating, maintaining, and overseeing the equipment. Include roles such as operators, technicians, and supervisors.

3. **Scope and Applicability:**

   Specify the cleanroom areas where the SOP is applicable. Clearly outline any restrictions or limitations regarding equipment use and access.

4. **Regulatory Compliance:**

   Ensure that the SOP aligns with relevant industry standards, regulations, and cleanroom classifications. Incorporate any legal or safety requirements into the procedure.

5. **Equipment Description:**

   Provide detailed information about the equipment, tool, or device covered by the SOP. Include the manufacturer's name, model, serial number, and any other pertinent details.

6. **Safety Precautions:**

   Clearly outline safety precautions that operators must follow when using the equipment. Include information on personal protective equipment (PPE), emergency shutdown procedures, and handling hazardous materials.

7. **Operating Procedures:**

Break down the operating steps into clear and concise instructions. Use a step-by-step format with detailed explanations. Include any specific settings, parameters, or configurations that need to be observed.

8. **Calibration and Maintenance:**

Detail the calibration and maintenance procedures for the equipment. Specify the frequency of calibration, routine maintenance tasks, and any required documentation.

9. **Troubleshooting:**

Include a troubleshooting section that addresses common issues, error messages, or malfunctions. Provide step-by-step guidance on identifying and resolving problems.

10. **Emergency Procedures:**

Outline emergency procedures related to the equipment. This may include shutdown protocols, evacuation steps, and contacts for technical support or maintenance.

11. **Documentation and Record Keeping:**

Specify the documentation requirements associated with the equipment. This may involve keeping logs, records of maintenance activities, or any other relevant documentation.

12. **Training Requirements:**

Identify the training requirements for personnel who will be operating or maintaining the equipment. Specify the qualifications or certifications needed.

13. **Review and Revision:**

Establish a schedule for regular reviews of the SOP. Clearly outline the process for revising the document to incorporate updates, improvements, or changes in equipment specifications.

14. **Approval Process:**

Define the approval process for the SOP. Specify who has the authority to approve, review, or make changes to the document.

15. **Distribution:**

> Clearly state how the SOP will be distributed to relevant personnel. Ensure that all individuals involved with the equipment have access to the latest version of the SOP.

16. **Implementation Plan:**

> Develop a plan for the implementation of the SOP. Communicate changes, conduct training sessions, and ensure that all users are familiar with the new procedures.

17. **Version Control:**

> Implement a version control system to track changes to the SOP. Clearly mark each version with a revision number and date.

18. **Record Retention:**

> Specify how long records associated with the equipment will be retained. Ensure compliance with any regulatory requirements regarding recordkeeping.

19. **Feedback Mechanism:**

> Establish a mechanism for receiving feedback on the SOP from users. Encourage personnel to report any issues, concerns, or suggestions for improvement.

20. **Continuous Improvement:**

> Foster a culture of continuous improvement. Regularly assess the effectiveness of the SOP and make updates as necessary to enhance efficiency, safety, and compliance.

Remember, the SOP should be written in a clear and accessible language, and it should serve as a comprehensive guide for anyone interacting with the equipment in the cleanroom environment. Regular reviews and updates are essential to ensure that the SOP remains current and effective.

## 14.2 What is Hazardous Gas Response Team (HGRT)?

A Hazardous Gas Response Team is a specialized team of individuals trained to respond to incidents involving hazardous gases. The primary goal of such a team is to mitigate and manage emergencies related to the release, leakage, or presence of dangerous gases in a facility or the environment. These teams are typically found in settings where the use, storage, or production of hazardous gases is a part of the operations, such as industrial facilities, laboratories, or cleanroom environments.

**Key aspects of a Hazardous Gas Response Team include:**

1. **Training and Expertise:**

   Team members undergo specialized training in handling hazardous gases, understanding their properties, and responding to emergencies. This training often includes knowledge of the specific gases used in the facility, their potential risks, and appropriate response measures.

2. **Emergency Response Planning:**

   The team develops and implements emergency response plans for incidents involving hazardous gases. This includes strategies for containment, evacuation procedures, and coordination with other emergency response teams or external agencies.

3. **Equipment and Gear:**

   Hazardous Gas Response Teams are equipped with specialized gear and equipment to safely respond to incidents. This may include personal protective equipment (PPE), gas detectors, containment devices, and tools for sealing or repairing leaks.

4. **Monitoring and Detection:**

   Team members are trained to use gas monitoring and detection equipment to assess the presence and concentration of hazardous gases. This allows them to identify potential risks and take appropriate actions.

5. **Coordination with Other Teams:**

   In larger facilities or those with multiple emergency response teams, the Hazardous Gas Response Team collaborates with other teams, such as the general emergency response team, fire brigade, or environmental health and safety team.

6. **Drills and Exercises:**

   Regular drills and exercises are conducted to ensure the team's preparedness for different scenarios. These simulations help team members practice their response strategies and improve coordination.

7. **Communication Protocols:**

> Effective communication is crucial during a hazardous gas incident. The team establishes clear communication protocols to coordinate actions internally and communicate with other relevant parties, including emergency services if necessary.

8. **Documentation and Reporting:**

> After responding to an incident, the Hazardous Gas Response Team is responsible for documenting the event, actions taken, and outcomes. This documentation is valuable for post-incident analysis and continuous improvement.

9. **Regulatory Compliance:**

> The team ensures compliance with relevant safety regulations, industry standards, and environmental regulations related to the handling and response to hazardous gases.

10. **Continuous Improvement:**

> Similar to other emergency response teams, the Hazardous Gas Response Team engages in continuous improvement efforts. This involves reviewing incidents, identifying areas for improvement, and updating response plans accordingly.

Having a dedicated Hazardous Gas Response Team is essential for maintaining a safe working environment, preventing accidents, and responding effectively in the event of a hazardous gas-related emergency.

## 14.3 Describe the HAZWOPE

HAZWOPER stands for Hazardous Waste Operations and Emergency Response. It refers to a set of guidelines and regulations established by the Occupational Safety and Health Administration (OSHA) in the United States. HAZWOPER outlines safety and health requirements for workers who are involved in handling, managing, and responding to hazardous waste operations or emergency situations.

**Key aspects of HAZWOPER include:**

1. **Training Requirements:**

> Workers involved in hazardous waste operations or emergency response must receive specific training based on their job roles. Training covers

areas such as hazard recognition, use of personal protective equipment (PPE), emergency response procedures, and other safety protocols.

2. **Hazardous Waste Operations:**

   HAZWOPER standards apply to workers engaged in activities related to the cleanup, treatment, storage, and disposal of hazardous waste. This includes personnel involved in the operation of treatment facilities, hazardous waste storage areas, and cleanup of hazardous waste sites.

3. **Emergency Response:**

   - The regulations also address workers who may be called upon to respond to emergencies involving hazardous substances. This includes scenarios such as spills, leaks, or accidents that require prompt and effective response to prevent harm to individuals and the environment.

4. **Levels of Training:**

   HAZWOPER training is categorized into different levels based on the roles and responsibilities of workers. These levels include First Responder Awareness Level, First Responder Operations Level, Hazardous Materials Technician Level, and Hazardous Materials Specialist Level.

5. **Safety Plans and Procedures:**

   Employers are required to develop and implement safety plans and procedures specific to their hazardous waste operations. These plans address aspects such as site safety, communication, decontamination procedures, and emergency response.

6. **Personal Protective Equipment (PPE):**

   HAZWOPER emphasizes the use of appropriate personal protective equipment to minimize exposure to hazardous substances. This may include protective clothing, respiratory protection, gloves, and other gear depending on the nature of the work.

7. **Medical Surveillance:**

   Employers are required to establish medical surveillance programs for workers involved in hazardous waste operations. This includes pre-employment medical examinations and regular health monitoring to identify and address any adverse health effects.

8. **Site Control Measures:**

    HAZWOPER addresses the need for site control measures to restrict access to hazardous waste areas. This includes the use of barricades, signs, and other measures to prevent unauthorized personnel from entering hazardous zones.

9. **Decontamination Procedures:**

    Procedures for the decontamination of workers and equipment are outlined in HAZWOPER. This is crucial to prevent the spread of hazardous substances and protect workers from exposure.

10. **Post-Emergency Response:**

    After an emergency response, there are provisions for post-emergency response activities, including cleanup, decontamination, and follow-up actions to ensure that the site is safe and free from hazardous substances.

HAZWOPER is designed to protect workers and the environment during activities involving hazardous waste. Compliance with HAZWOPER standards is crucial for organizations and individuals engaged in these operations to prevent accidents, injuries, and environmental contamination.

## 14.4 How to become a Qualified Cleanroom Electrical Worker?

Becoming a qualified electrical worker involves a combination of education, training, and experience. Here are the general steps to become a qualified electrical worker:

1. **Educational Requirements:**

    Obtain a high school diploma or equivalent. High school coursework in mathematics, physics, and technology can be beneficial.

2. **Pursue Electrical Education:**

    Consider enrolling in a post-secondary electrical education program. This could be a certificate, diploma, or degree program in electrical technology or a related field. Look for programs accredited by relevant authorities or professional organizations.

3. **Apprenticeship Training:**

    Many aspiring electricians enter apprenticeship programs to gain practical, on-the-job training. Apprenticeships typically last four to five years and involve a combination of classroom instruction and hands-on experience. Seek apprenticeship opportunities through local trade unions, electrical contractors, or vocational schools.

4. **Licensing and Certification:**

In many regions, electricians are required to be licensed or certified. Check with the local regulatory authorities to understand the specific licensing requirements in your area. This may involve passing an exam and fulfilling a certain number of work hours.

5. **Gain Practical Experience:**

Work as an apprentice to gain hands-on experience in electrical installations, maintenance, and repairs. During this time, you'll learn to read blueprints, install wiring, troubleshoot electrical problems, and adhere to safety regulations.

6. **Specialization (Optional):**

Consider specializing in a particular area of electrical work, such as residential, commercial, industrial, or renewable energy. Specialization can enhance your skills and open up more opportunities in specific sectors.

7. **Continuing Education:**

Stay updated on the latest developments in electrical technology and safety regulations. Consider taking additional courses or workshops to expand your knowledge and skills.

8. **Obtain Additional Certifications (Optional):**

Depending on your career goals and the specific requirements in your region, you may choose to obtain additional certifications, such as becoming a certified electrician in a specialized area.

9. **Networking:**

Build a professional network by connecting with other electricians, contractors, and professionals in the field. Networking can provide valuable insights, job opportunities, and mentorship.

10. **Adherence to Safety Regulations:**

Always prioritize safety in your work. Understand and adhere to electrical codes and safety regulations. Continuous awareness of safety measures is crucial in the electrical industry.

11. **Professional Associations:**

> Consider joining professional associations for electricians. These organizations may provide resources, support, and networking opportunities.

12. **Update Your Resume:**

> Keep your resume updated with your education, training, and work experience. Highlight any certifications or specializations you have acquired.

Remember that the specific requirements and steps may vary depending on your location. It's important to research and understand the regulations and licensing procedures in your region to ensure compliance.

# Chapter 15: Collaborating with Scientific Projects

## 15.1 Working with Principal Investigators

Cooperating with Principal Investigators (PIs) involved in scientific projects is crucial for a cleanroom manager to ensure the successful integration of cleanroom facilities into their research activities. Here's a guide on how a cleanroom manager can effectively collaborate with scientific project PIs:

**1. Initial Collaboration:**

- **Introduction and Familiarization:**

  - Initiate a meeting with the PI to introduce yourself and the cleanroom facilities.

  - Provide a comprehensive overview of the cleanroom's capabilities, equipment, and safety protocols.

- **Understanding Research Goals:**

  - Gain a clear understanding of the PI's research goals and objectives.

  - Discuss the specific cleanroom requirements for their experiments and projects.

**2. Customized Cleanroom Access:**

- **Tailored Training Programs:**

  - Develop customized training programs for cleanroom users based on the nature of the PI's experiments.

  - Ensure that users are well-versed in cleanroom protocols and safety measures.

- **Flexible Scheduling:**

  - Collaborate with the PI to establish flexible scheduling arrangements that align with their experimental timelines.

  - Accommodate specific time slots for critical experiments.

**3. Technical Support:**

- **Collaborative Experiment Design:**

  - Work closely with PIs during the experimental design phase to address any technical considerations related to cleanroom tools and equipment.

- Provide technical expertise to optimize experimental setups.

- **Troubleshooting Assistance:**

  - Offer ongoing technical support and troubleshooting assistance for any issues related to cleanroom equipment.

  - Ensure a rapid response to maintenance and technical challenges.

## 4. Research Project Integration:

- **Integration Planning:**

  - Collaborate on the integration of cleanroom activities into the broader research project.

  - Identify potential synergies with other cleanroom users or projects.

- **Cross-Functional Collaboration:**

  - Facilitate collaboration between different PIs and researchers using the cleanroom.

  - Encourage interdisciplinary discussions to foster innovation.

## 5. Communication and Feedback:

- **Regular Meetings:**

  - Schedule regular meetings with PIs to discuss ongoing and upcoming experiments.

  - Address any concerns or feedback related to cleanroom operations.

- **Documentation and Reporting:**

  - Maintain clear documentation of cleanroom usage for each research project.

  - Provide periodic reports on usage statistics, equipment status, and any incidents.

## 6. Safety and Compliance:

- **Safety Protocols:**

  - Collaborate with PIs to ensure that all safety protocols are followed during experiments.

  - Conduct regular safety reviews and training sessions.

- **Regulatory Compliance:**

    - Assist PIs in understanding and adhering to regulatory requirements relevant to their experiments.

    - Provide guidance on documentation and reporting for regulatory compliance.

## 7. Resource Allocation:

- **Optimizing Tool Usage:**

    - Collaborate with PIs to optimize the use of cleanroom tools and equipment.

    - Implement efficient scheduling to accommodate varying levels of demand.

- **Budgetary Collaboration:**

    - Work with PIs to allocate budgetary resources effectively for cleanroom usage.

    - Assist in grant applications related to cleanroom access.

## 8. Continuous Improvement:

- **Feedback Mechanism:**

    - Establish a feedback mechanism to gather input from PIs on their experiences with cleanroom facilities.

    - Use feedback to identify areas for improvement and implement necessary changes.

- **Technology Updates:**

    - Keep PIs informed about updates to cleanroom technologies or the addition of new equipment.

    - Discuss potential upgrades to enhance capabilities.

**Conclusion:**

A collaborative and proactive approach to working with scientific project PIs is essential for a cleanroom manager. By understanding the unique requirements of each research project, providing tailored support, and fostering open communication, the cleanroom manager can contribute significantly to the success of scientific endeavors conducted within the cleanroom

# 15.2 writing equipment proposals

Writing equipment proposals for acquiring new tools for the cleanroom involves careful planning, clear communication, and a focus on how the new equipment will enhance the cleanroom's capabilities. Here is a step-by-step guide on how to write an effective equipment proposal:

## 1. Identify Equipment Needs:

### Conduct Needs Assessment:

- Assess the current cleanroom capabilities and identify specific needs or gaps that the new equipment can address.

- Consult with cleanroom users, researchers, and other stakeholders to gather input on required tools.

## 2. Research Equipment Options:

### Market Research:

- Investigate available equipment options from reputable suppliers and manufacturers.

- Compare specifications, features, and prices to determine the most suitable equipment for the cleanroom's requirements.

## 3. Define Project Scope:

### Clearly Outline Objectives:

- Clearly state the objectives of acquiring the new equipment.

- Define how the equipment aligns with the cleanroom's goals and enhances research capabilities.

## 4. Develop a Proposal Outline:

- **Introduction:**

  - Provide an overview of the cleanroom and its current capabilities.

  - Introduce the need for the new equipment and its significance.

- **Project Description:**

  - Describe the specific equipment being proposed.

  - Highlight key features and capabilities that make the equipment suitable for the cleanroom.

- **Budget and Funding:**
  - Present a detailed budget for acquiring and installing the equipment.
  - Clearly outline the funding sources or grant applications that will cover the costs.
- **Timeline:**
  - Provide a timeline for the acquisition, delivery, installation, and commissioning of the equipment.
  - Indicate any dependencies or milestones related to the project.
- **Benefits and Impact:**
  - Outline the anticipated benefits of acquiring the new equipment.
  - Explain how the equipment will positively impact research outcomes and cleanroom efficiency.

## 5. Budget Justification:

- **Breakdown of Costs:**
  - Break down the budget into categories such as equipment purchase, installation, training, and any additional costs.
  - Justify each cost item with a brief explanation.
- **Cost-Benefit Analysis:**
  - Provide a cost-benefit analysis that demonstrates the value the new equipment will bring to the cleanroom.
  - Compare the long-term benefits to the initial investment.

## 6. User Training and Support:

- **Training Plan:**
  - Outline a plan for training cleanroom users on how to operate the new equipment.
  - Detail any additional support or resources needed for ongoing maintenance and training.

## 7. Collaboration and Integration:

- **Integration with Existing Infrastructure:**

    - Explain how the new equipment will integrate with existing cleanroom infrastructure.

    - Highlight any collaboration opportunities with other research projects.

- **Cross-Functional Benefits:**

    - Emphasize how the new equipment can benefit multiple research areas and projects within the cleanroom.

## 8. Risk Assessment and Mitigation:

- **Identify Risks:**

    - Acknowledge potential risks or challenges associated with acquiring and implementing the new equipment.

    - Include a risk assessment that considers factors such as technical challenges, delays, or budget overruns.

- **Mitigation Strategies:**

    - Propose mitigation strategies to address identified risks.

    - Demonstrate preparedness and contingency plans to minimize potential negative impacts.

## 9. Appendix:

- **Supporting Documentation:**

    - Include any supporting documentation, such as quotations, technical specifications, or letters of support.

    - Attach relevant documents that strengthen the proposal's credibility.

## 10. Review and Finalize:

- **Internal Review:**

    - Conduct an internal review of the proposal before submission.

    - Seek feedback from relevant stakeholders to ensure clarity and completeness.

- **Finalize Proposal:**

    - Make necessary revisions based on feedback.

    - Ensure that the final proposal is well-organized, articulate, and aligned with the cleanroom's strategic objectives.

## 11. Submission:

- **Submit to Decision-Makers:**

    - Submit the finalized proposal to decision-makers, such as cleanroom management, research committees, or funding bodies.

    - Follow any specific submission guidelines provided by the reviewing entities.

## 12. Follow-Up:

- **Engage in Communication:**

    - Engage in open communication with decision-makers during the review process.

    - Be prepared to provide additional information or address questions that may arise.

By following this comprehensive guide, a cleanroom manager can create a well-structured equipment proposal that effectively communicates the need for new tools and justifies the investment in enhancing cleanroom capabilities.

## 15.3 Conducting cleanroom outreach activities (tours, discussions, etc.)

Helping cleanroom staff with cleanroom outreach activities involves planning and executing initiatives that showcase the cleanroom's significance, technology, and research. Here's a step-by-step guide on how to facilitate cleanroom outreach activities such as tours and discussions:

## 1. Define Outreach Goals:

- Clearly define the objectives of the outreach activities.

- Determine whether the goal is to educate students, engage with the community, or foster collaboration with industry partners.

## 2. Identify Target Audience:

- Identify the target audience for the outreach activities.

- Tailor the content and format of the outreach to suit the needs and interests of the audience, whether they are students, professionals, or community members.

## 3. Coordinate with Cleanroom Management:

- Collaborate with cleanroom management to gain support and approval for outreach activities.
- Ensure that the proposed activities align with the cleanroom's schedule, safety protocols, and ongoing research activities.

## 4. Develop Outreach Materials:

- Create informative and engaging materials that explain the cleanroom's purpose, technology, and research areas.
- Design brochures, pamphlets, or multimedia presentations that are visually appealing and easy to understand.

## 5. Organize Cleanroom Tours:

- Plan and schedule cleanroom tours for different groups.
- Ensure that tours are guided by knowledgeable staff who can explain the cleanroom's functions, safety measures, and ongoing projects.

## 6. Interactive Demonstrations:

- Include interactive demonstrations to illustrate cleanroom processes and technologies.
- Engage the audience by allowing them to observe or participate in specific cleanroom activities.

## 7. Q&A Sessions:

- Incorporate question-and-answer sessions to address the audience's curiosity.
- Prepare cleanroom staff to answer common questions about cleanroom technology, safety, and research.

## 8. Collaborate with Educational Institutions:

- Reach out to local schools, colleges, and universities to organize educational outreach programs.
- Offer tailored presentations or workshops that align with educational curricula.

**9. Industry Collaborations:**

- Collaborate with industry partners for specialized cleanroom outreach events.
- Highlight industry applications of cleanroom technology and research.

**10. Social Media and Online Platforms:**

- Leverage social media and online platforms to extend the reach of cleanroom outreach.
- Share informative content, virtual tours, or live sessions to engage a broader audience.

**11. Engage with Community Groups:**

- Partner with community organizations to organize cleanroom discussions or presentations.
- Emphasize the cleanroom's impact on local communities and potential collaboration opportunities.

**12. Evaluate and Gather Feedback:**

- Collect feedback from participants to assess the effectiveness of outreach activities.
- Use feedback to improve future outreach initiatives and address any areas of improvement.

**13. Maintain Communication Channels:**

- Establish and maintain communication channels for ongoing engagement.
- Keep the audience informed about cleanroom updates, research milestones, or upcoming events.

**14. Highlight Collaborative Projects:**

- Showcase collaborative projects between the cleanroom and external partners.
- Illustrate the real-world applications and impact of cleanroom research.

**15. Document Outreach Success Stories:**

- Document success stories or testimonials resulting from cleanroom outreach.
- Use success stories to showcase the positive outcomes and benefits of engaging with the cleanroom.

By following these steps, cleanroom staff can effectively plan and execute outreach activities that educate, engage, and create a positive impact on the community, students, and industry partners.

# Chapter 16: Coordinating Facilities Maintenance

## 16.1 Coordinating Facilities Modification

Coordinating facilities modifications as a cleanroom manager involves working closely with various groups to ensure seamless integration of changes while minimizing disruptions to cleanroom operations. Here's a step-by-step guide on how to effectively coordinate facilities modifications:

**1. Assessment and Planning:**

- **Identify Modification Needs:**

    - Conduct a thorough assessment to identify the specific needs for facilities modifications.

    - Consider input from cleanroom users, maintenance records, and any regulatory requirements.

- **Define Objectives:**

    Clearly define the objectives of the modifications, whether it's upgrading equipment, expanding cleanroom space, or enhancing safety features.

- **Budgetary Considerations:**

    - Work with relevant groups, such as finance and administration, to establish a budget for the proposed modifications.

    - Ensure alignment with overall cleanroom and organizational budgets.

**2. Collaboration with Stakeholders:**

- **Engage Cleanroom Users:**

    - Communicate with cleanroom users to understand their specific needs and concerns.

    - Gather input on the potential impact of modifications on ongoing projects.

- **Consult with Facilities/Operations:**

    - Collaborate with facilities and operations teams to assess the feasibility of proposed modifications.

    - Discuss technical aspects, construction requirements, and timelines.

- **Involve Administration:**
    - Keep administrative teams informed about the proposed modifications.
    - Address administrative considerations such as procurement, contracts, and approvals.

## 3. Regulatory Compliance:

- **Understand Regulations:**
    - Ensure that all proposed modifications comply with relevant cleanroom regulations and standards.
    - Collaborate with regulatory affairs teams to navigate compliance requirements.

- **Documentation:**
    - Maintain thorough documentation of proposed modifications, including regulatory approvals and permits.
    - Provide necessary documentation for audits or inspections.

## 4. Project Management:

- **Appoint Project Manager:**
    - Designate a project manager responsible for overseeing the modifications.
    - Ensure the project manager has the necessary skills and resources to manage the project effectively.

- **Timeline Development:**
    - Develop a detailed timeline for the modifications, considering critical milestones and deadlines.
    - Communicate the timeline to all stakeholders to manage expectations.

- **Regular Updates:**
    - Provide regular updates to cleanroom users, facilities teams, and administration throughout the project.
    - Address any unexpected challenges promptly and transparently.

**5. Communication and Outreach:**

- **Internal Communication:**
    - Maintain open and clear communication channels within the cleanroom community.
    - Use email, newsletters, or internal communication platforms to keep everyone informed.
- **Outreach Activities:**
    - Organize discussions or forums to address concerns and questions from cleanroom staff.
    - Conduct tours or presentations to showcase the planned modifications.

**6. Risk Mitigation:**

- **Risk Assessment:**
    - Conduct a comprehensive risk assessment for the modifications.
    - Identify potential risks and develop mitigation strategies to minimize their impact.
- **Contingency Plans:**
    - Develop contingency plans for potential disruptions to cleanroom operations during the modifications.
    - Ensure there are backup plans in place for critical activities.

**7. Post-Modification Evaluation:**

- **Feedback Collection:**
    - Collect feedback from cleanroom users after the modifications are complete.
    - Use feedback to assess the success of the modifications and identify areas for improvement.
- **Continuous Improvement:**
    - Implement continuous improvement measures based on lessons learned from the modification process.
    - Document best practices for future facility modifications.

## 8. Documentation and Reporting:

- **Final Report:**
    - Compile a final report summarizing the modifications, including costs, timelines, and outcomes.
    - Share the report with relevant stakeholders for transparency and accountability.

- **Archiving:**
    - Maintain a comprehensive archive of all documentation related to the modifications.
    - Ensure accessibility for future reference or audits.

## 9. Celebrating Success:

**Recognition:**

- Acknowledge the efforts of all involved parties in the successful completion of the modifications.
- Celebrate achievements and recognize contributions from various teams.

By following these steps, a cleanroom manager can effectively coordinate facilities modifications, fostering collaboration among different groups and ensuring a smooth and successful implementation of changes within the cleanroom environment.

## 16.2 Resolving Facilities Problems

Coordinating efforts to resolve facilities problems with equipment in a cleanroom requires a systematic and collaborative approach. As a cleanroom manager, you play a crucial role in ensuring that facilities problems are addressed promptly to maintain optimal operational conditions. Here's a step-by-step guide:

## 1. Immediate Assessment:

- **Receive Incident Report:**
    - Promptly receive incident reports regarding facilities problems with equipment.
    - Encourage cleanroom users to report issues as soon as they are identified.

- **Preliminary Investigation:**
  - Conduct a preliminary assessment to understand the nature and severity of the problem.
  - Verify the reported issue and gather relevant information.

## 2. Engage Relevant Teams:

- **Coordinate with Technical Teams:**
  - Engage with technical teams responsible for maintaining and servicing equipment.
  - Share information about the reported problem and request their expertise.
- **Involve Facilities Management:**
  - Collaborate with facilities management to assess any potential impact on the cleanroom environment.
  - Discuss solutions and resources needed for resolution.

## 3. Prioritization:

- **Evaluate Severity:**
  - Assess the severity of the facilities problem and its impact on cleanroom operations.
  - Prioritize issues based on their criticality and potential for disruption.
- **Emergency Response:**
  - If the problem poses an immediate threat, initiate emergency response protocols.
  - Ensure the safety of cleanroom users and take appropriate actions.

## 4. Communication:

- **Notify Cleanroom Users:**
  - Keep cleanroom users informed about the status of the reported problem.
  - Provide updates on the resolution process and any temporary measures in place.

- **Collaborative Communication:**

  - Facilitate communication between technical teams, facilities management, and cleanroom users.

  - Establish a collaborative environment for problem-solving.

## 5. Root Cause Analysis:

- **Technical Evaluation:**

  - Work closely with technical teams to conduct a detailed root cause analysis.

  - Identify the underlying issues leading to equipment problems.

- **Document Findings:**

  - Document the findings of the analysis, including any identified failures or maintenance issues.

  - Use this documentation for future reference and preventive measures.

## 6. Resolution Planning:

- **Develop Action Plan:**

  - Collaborate with technical teams to develop a comprehensive action plan for resolution.

  - Clearly outline steps, timelines, and resources required for each phase.

- **Consider User Impact:**

  - Assess the potential impact of the resolution process on cleanroom users.

  - Minimize disruptions by scheduling maintenance during low-usage periods if possible.

## 7. Resource Allocation:

- **Allocate Resources:**

  - Ensure that necessary resources, including spare parts and specialized tools, are available for the resolution.

  - Coordinate with procurement or relevant departments to expedite resource allocation.

- **Facilities Support:**

    - Work with facilities management to ensure that the cleanroom environment supports the resolution process.

    - Provide any necessary accommodations for maintenance activities.

## 8. Monitoring and Follow-Up:

- **Continuous Monitoring:**

    - Monitor the progress of the resolution activities.

    - Address any unforeseen challenges promptly and adjust the action plan if needed.

- **User Feedback:**

    - Seek feedback from cleanroom users regarding the effectiveness of the resolution.

    - Use feedback to improve future response strategies.

## 9. Documentation and Reporting:

- **Maintain Records:**

    - Keep detailed records of the reported problem, root cause analysis, and resolution process.

    - Document the entire incident for audit and analysis purposes.

- **Report to Leadership:**

    - Provide leadership with a comprehensive report on the incident, its resolution, and any preventive measures recommended.

    - Ensure transparency and accountability.

## 10. Preventive Measures:

- **Implement Preventive Measures:**

    - Work with technical teams to implement preventive measures to avoid similar issues in the future.

    - Update maintenance schedules or procedures if necessary.

- **Training and Awareness:**
    - Conduct training sessions for cleanroom users on proper equipment usage and early issue reporting.
    - Enhance user awareness of the importance of preventive maintenance.

By coordinating these efforts, a cleanroom manager can effectively address facilities problems with equipment, minimize downtime, and contribute to the overall reliability and functionality of the cleanroom environment.

## 16.3 Types of Maintenance in Cleanrooms

As a cleanroom manager, you'll oversee various types of maintenance to ensure the optimal performance of cleanroom tools, equipment, and devices. Here are the main types of maintenance and brief definitions along with examples:

**1. Preventive Maintenance (PM):**

- **Definition:** Planned and routine maintenance activities performed to prevent equipment failure or deterioration.

- **Examples:**
    - **Tool Calibration:** Regular calibration of metrology tools such as particle counters.
    - **Filter Replacement:** Scheduled replacement of HEPA filters in laminar flow hoods.
    - **Cleaning:** Regular cleaning of equipment surfaces to prevent particle contamination.

**2. Corrective Maintenance:**

- **Definition:** Unscheduled maintenance performed to correct equipment failures or malfunctions.

- **Examples:**
    - **Motor Replacement:** Replacing a malfunctioning motor in a vacuum pump.
    - **Sensor Calibration:** Adjusting and calibrating sensors that are providing inaccurate readings.
    - **Circuit Board Repair:** Fixing a malfunctioning circuit board in a deposition tool.

**3. Predictive Maintenance:**

- **Definition:** Using data and analytics to predict when equipment is likely to fail, allowing for timely maintenance.

- **Examples:**

    - **Vibration Analysis:** Monitoring vibrations in rotating equipment to predict bearing failures.

    - **Thermal Imaging:** Detecting abnormal temperature patterns in electrical components.

    - **Fluid Analysis:** Analyzing the condition of lubricating fluids to predict equipment wear.

**4. Condition-Based Maintenance:**

- **Definition:** Performing maintenance based on the actual condition of equipment, as indicated by monitoring parameters.

- **Examples:**

    - **Pressure Checks:** Regularly monitoring pressure levels in gas delivery systems.

    - **Temperature Monitoring:** Observing temperature variations in critical components.

    - **Flow Rate Monitoring:** Checking and adjusting flow rates in liquid handling systems.

**5. Reactive Maintenance:**

- **Definition:** Addressing maintenance needs only when equipment failure occurs, often due to budget constraints or lack of planning.

- **Examples:**

    - **Emergency Repairs:** Fixing a sudden failure in a critical tool that hinders production.

    - **Replacement of Failed Components:** Swapping out failed components on an as-needed basis.

    - **Emergency Filter Change:** Changing filters when a sudden increase in particle counts is observed.

**6. Reliability-Centered Maintenance (RCM):**

- **Definition:** A systematic approach to determine the most effective maintenance strategy for each piece of equipment.

- **Examples:**

    - **Failure Mode and Effects Analysis (FMEA):** Identifying potential failure modes and their consequences.

    - **Criticality Analysis:** Determining the criticality of each piece of equipment for the overall process.

    - **Optimizing Maintenance Intervals:** Adjusting maintenance schedules based on equipment criticality.

## 7. Total Productive Maintenance (TPM):

- **Definition:** An approach that integrates maintenance into the daily operations of the entire organization to maximize equipment effectiveness.

- **Examples:**

    - **Employee Training:** Training operators to perform basic equipment checks and routine maintenance.

    - **Autonomous Maintenance:** Empowering operators to take responsibility for equipment cleanliness and basic maintenance.

    - **Early Equipment Management:** Involving maintenance teams in the design and installation of new equipment.

## 8. Calibration and Verification:

- **Definition:** Ensuring that measurement and control devices provide accurate and reliable results.

- **Examples:**

    - **Metrology Tools:** Calibrating particle counters, temperature sensors, and pressure gauges.

    - **Weighing Scales:** Regularly verifying the accuracy of weighing scales used for material measurements.

    - **Flow Meters:** Calibrating flow meters to ensure precise measurement of fluid or gas flow.

**9. Lubrication:**

- **Definition:** Applying the appropriate lubricants to equipment to reduce friction and prevent wear.

- **Examples:**

    - **Vacuum Pump Lubrication:** Regularly lubricating vacuum pump components to ensure smooth operation.

    - **Bearing Lubrication:** Applying lubricants to bearings in rotating equipment.

    - **Linear Motion Systems:** Lubricating linear slides and rails to prevent friction-related issues.

**10. Cleanroom Environment Monitoring:**

- **Definition:** Regular monitoring of environmental conditions to ensure compliance with cleanroom standards.

- **Examples:**

    - **Particle Monitoring:** Continuous monitoring of airborne particle counts.

    - **Temperature and Humidity Monitoring:** Ensuring temperature and humidity levels are within specified ranges.

    - **Pressure Differentials:** Monitoring pressure differentials to prevent contamination.

These maintenance types contribute to the overall reliability, performance, and cleanliness of a cleanroom environment. Implementing a comprehensive maintenance strategy helps minimize downtime, extend equipment lifespan, and maintain the integrity of cleanroom processes.

## 16.4 Scheduling Periodic Maintenance

Scheduling periodic maintenance for chillers, pumps, and other critical equipment in a cleanroom involves a systematic approach to ensure optimal performance and minimize downtime. Here's a step-by-step guide:

1. **Asset Inventory:**

    - Maintain a comprehensive inventory of all equipment in the cleanroom, including chillers, pumps, HVAC systems, and other critical components.

    - Document specifications, manufacturer details, installation dates, and historical maintenance records for each asset.

2. **Risk Assessment:**

- Conduct a risk assessment to identify critical equipment that may impact cleanroom operations if it fails.
- Prioritize equipment based on criticality and potential consequences of failure.

3. **Manufacturer Guidelines:**

- Refer to manufacturer guidelines and specifications for recommended maintenance schedules.
- Follow the manufacturer's recommendations for routine checks, lubrication, and replacement of consumables.

4. **Regulatory Compliance:**

- Ensure compliance with industry standards and regulations governing cleanroom operations.
- Understand specific requirements related to the maintenance of equipment in controlled environments.

5. **Maintenance Calendar:**

- Develop a centralized maintenance calendar that outlines scheduled activities for each piece of equipment.
- Clearly define the frequency of checks, inspections, and major overhauls based on equipment type and criticality.

6. **Preventive Maintenance Plans:**

- Implement preventive maintenance plans for chillers, pumps, and other equipment.
- Define routine tasks such as cleaning, lubrication, filter replacement, and calibration.

7. **CMMS (Computerized Maintenance Management System):**

- Utilize a CMMS to automate and streamline the scheduling and tracking of maintenance activities.
- Input equipment details, set up recurring maintenance tasks, and generate work orders as needed.

8. **Collaboration with Operations Teams:**

- Collaborate with cleanroom operators and production teams to identify suitable time slots for scheduled maintenance that minimize disruption to ongoing processes.

- Communicate maintenance schedules well in advance to relevant stakeholders.

9. **Critical Downtime Planning:**

- Plan for critical downtime during maintenance activities by coordinating with production schedules.

- Implement strategies to mitigate the impact of downtime on overall cleanroom productivity.

10. **Documentation and Reporting:**

- Keep detailed records of all maintenance activities, including dates, tasks performed, and any identified issues.

- Generate regular reports summarizing equipment performance, completed maintenance, and upcoming schedules.

11. **Training and Certification:**

- Ensure that maintenance personnel are adequately trained and certified to perform tasks related to specific equipment.

- Stay informed about advancements in maintenance technologies and industry best practices.

12. **Continuous Improvement:**

- Regularly review and assess the effectiveness of the maintenance schedule.

- Implement continuous improvement measures based on feedback, equipment performance data, and emerging technologies.

By following these steps, cleanroom managers can establish an effective and proactive approach to scheduling periodic maintenance, ensuring the reliability and longevity of critical equipment in the cleanroom environment.

## 16.5 Inventory Management for Continuous Operation

Coordinating supplies and maintaining an efficient inventory system is crucial for the continuous operation of a cleanroom. Here's a guide for a cleanroom manager:

1. **Inventory Management System:**

    - Implement a robust inventory management system that tracks supplies, consumables, and critical gases used in the cleanroom.

    - Utilize a computerized system or specialized software to monitor inventory levels, order history, and consumption patterns.

2. **Categorization of Supplies:**

    - Categorize supplies based on criticality and usage frequency.

    - Classify items as consumables, process gases, chemicals, or other essential materials.

3. **Par Levels:**

    - Establish par levels for each category of supplies to ensure that there is always an adequate quantity on hand.

    - Par levels should be determined based on historical consumption patterns and lead times for ordering.

4. **Supplier Relationships:**

    - Cultivate strong relationships with reliable suppliers for cleanroom-specific supplies, gases, and consumables.

    - Negotiate contracts and agreements to ensure a steady and timely supply of materials.

5. **Automated Reordering:**

    - Implement automated reordering systems linked to the inventory management system.

    - Set up alerts or triggers to automatically generate purchase orders when inventory levels drop below predetermined thresholds.

6. **Lead Time Considerations:**

    - Factor in lead times for ordering supplies, especially for items with longer procurement timelines.

    - Plan ahead for seasonal variations, supplier holidays, or any other factors that may affect delivery times.

7. **Regular Audits:**

   - Conduct regular audits of the inventory to verify physical stock levels against the recorded data.

   - Identify and investigate any discrepancies and take corrective actions promptly.

8. **Storage Conditions:**

   - Ensure proper storage conditions for supplies, especially for sensitive materials.

   - Monitor temperature, humidity, and other environmental factors that may impact the quality of stored items.

9. **Communication with Operations Teams:**

   - Maintain open communication with cleanroom operators and production teams to understand upcoming projects, experiments, or changes in demand.

   - Collaborate on forecasting to anticipate variations in supply requirements.

10. **Emergency Preparedness:**

    - Develop contingency plans for unexpected supply chain disruptions or emergencies.

    - Establish alternative suppliers or backup plans to mitigate risks to continuous cleanroom operation.

11. **Training for Personnel:**

    - Train cleanroom personnel on proper handling, storage, and tracking of supplies.

    - Implement a system for requesting additional supplies or reporting shortages.

12. **Regular Review and Optimization:**

    - Regularly review the inventory management processes and identify opportunities for optimization.

    - Utilize feedback from cleanroom staff to improve efficiency and responsiveness in the supply chain.

By effectively coordinating supplies and maintaining a well-organized inventory system, cleanroom managers can ensure the seamless and continuous operation of the cleanroom environment.

# Chapter 17: Communication and Updates

## 17.1 Providing Regular Updates to Staff/Users

Providing regular updates and communicating changes about cleanroom operations to scientific groups, staff/users, and user offices is crucial for maintaining transparency, efficiency, and safety. Here's a comprehensive guide on how to effectively handle this aspect:

1. **Establish Communication Protocols:**

   Define clear communication protocols and channels for disseminating information. This may include email newsletters, official announcements, dedicated communication platforms, or regular meetings.

2. **Regular Newsletters:**

   - Send out regular newsletters summarizing important updates, policy changes, and upcoming events within the cleanroom.

   - Include information on maintenance schedules, equipment availability, and any notable achievements or improvements.

3. **User Meetings:**

   - Conduct periodic user meetings to discuss operational changes, address concerns, and gather feedback.

   - Provide a platform for users to share their experiences and recommendations for improving cleanroom operations.

4. **Online Platforms:**

   - Utilize online platforms such as intranet, shared drives, or project management tools to host relevant documents, manuals, and updates.

   - Encourage users to check these platforms for the latest information.

5. **Scheduled Announcements:**

   - Schedule regular announcements or updates to ensure a consistent flow of information.

   - Highlight any changes in operating procedures, safety protocols, or access policies.

6. **Emergency Notifications:**

   - Establish a system for emergency notifications to quickly communicate critical information in urgent situations.
   - Ensure that users are aware of emergency protocols and how to receive real-time updates.

7. **Feedback Mechanism:**

   - Implement a feedback mechanism to allow users to submit their comments, concerns, or suggestions.
   - Actively address feedback and communicate improvements based on user input.

8. **Training Sessions:**

   - Conduct regular training sessions for new and existing users to familiarize them with cleanroom procedures and any recent updates.
   - Provide refresher courses on safety protocols and best practices.

9. **Collaboration with User Office:**

   - Collaborate closely with the cleanroom user office to ensure alignment in communication strategies.
   - Coordinate efforts to maintain a consistent and accurate flow of information.

10. **Documentation:**

    - Maintain up-to-date documentation on cleanroom policies, user guidelines, and standard operating procedures.
    - Ensure that all documentation is easily accessible to users.

11. **Communication Liaison:**

    - Designate a communication liaison or point of contact for users to direct inquiries or seek clarification.
    - Establish an efficient system for users to reach out with questions or concerns.

12. **Calendar of Events:**

- Create and share a calendar of events, including scheduled maintenance, workshops, and training sessions.

- Include details on how users can participate in or benefit from these events.

13. **Adaptive Communication:**

- Tailor communication strategies to the preferences of the user community, considering factors such as language, accessibility, and preferred communication channels.

14. **Celebrating Achievements:**

- Celebrate and communicate achievements, milestones, or successful projects within the cleanroom.

- Recognize the contributions of users and staff members.

By implementing these strategies, cleanroom managers can establish effective communication practices that foster collaboration, transparency, and a positive working environment within the cleanroom facility.

## 17.2 Cleanroom Contamination Levels

Cleanroom contamination levels are classified based on the maximum allowable number of airborne particles and microbes within a specified volume of air. These classifications are defined by international standards, such as ISO (International Organization for Standardization) and US Federal Standard 209E (no longer in use but still referenced). The ISO standard is now more widely accepted globally.

The contamination levels are typically expressed in terms of the maximum allowable particles per cubic meter of air for particles of a specific size range. Cleanrooms are classified into different classes, each with its own defined particle count limits. The two most commonly referenced standards are ISO 14644-1 and the older Federal Standard 209E.

Here's an overview of cleanroom contamination levels based on ISO 14644-1:

1. **ISO Class 1:**

- Maximum allowable particles: 10 particles/m³ for particles ≥0.1 µm

- This is the highest cleanliness level and is typically required in critical environments such as semiconductor manufacturing.

2. **ISO Class 2:**

   - Maximum allowable particles: 100 particles/m³ for particles ≥0.1 µm

   - Found in environments requiring extremely low levels of contamination, such as advanced optics manufacturing.

3. **ISO Class 3:**

   - Maximum allowable particles: 1,000 particles/m³ for particles ≥0.1 µm

   - Commonly used in semiconductor cleanrooms and pharmaceutical manufacturing.

4. **ISO Class 4:**

   - Maximum allowable particles: 10,000 particles/m³ for particles ≥0.1 µm

   - Suitable for less critical manufacturing processes, such as some medical device manufacturing.

5. **ISO Class 5:**

   - Maximum allowable particles: 100,000 particles/m³ for particles ≥0.1 µm

   - Often used in pharmaceutical manufacturing, laboratories, and electronics assembly.

6. **ISO Class 6:**

   - Maximum allowable particles: 1,000,000 particles/m³ for particles ≥0.1 µm

   - Suitable for applications where a controlled environment is essential but with less stringent requirements.

The classification continues in a similar manner with increasing allowable particle counts for each class. It's important to note that the classification is based on the count of particles of a specified size range (≥0.1 µm for ISO classes).

Additionally, microbial contamination is measured in Colony-Forming Units (CFU) per cubic meter. Cleanrooms may have additional classifications for microbial contamination, such as "Aseptic" for environments with the most stringent requirements for maintaining sterility.

These classifications are essential for industries where even minute particles can adversely affect the quality and performance of products, such as semiconductor manufacturing, pharmaceuticals, biotechnology, and aerospace.

## 17.3 ISO Classification of Cleanrooms?

The ISO classification of cleanrooms is defined in ISO 14644-1. This standard establishes the maximum allowable particle counts for airborne particles in cleanrooms and associated controlled environments. The classification is based on the number of particles present in the air per cubic meter at a specified particle size.

Here are the ISO cleanroom classes along with their corresponding maximum allowable particle counts:

1. **ISO Class 1:**

   - Maximum allowable particles/m³: 10 at 0.1 µm, 2 at 0.2 µm, 0 at 0.3 µm, 0 at 0.5 µm, 0 at 1 µm, 0 at 5 µm, and 0 at 10 µm.

2. **ISO Class 2:**

   - Maximum allowable particles/m³: 100 at 0.1 µm, 24 at 0.2 µm, 10 at 0.3 µm, 4 at 0.5 µm, 1 at 1 µm, 0 at 5 µm, and 0 at 10 µm.

3. **ISO Class 3:**

   - Maximum allowable particles/m³: 1,000 at 0.1 µm, 237 at 0.2 µm, 102 at 0.3 µm, 35 at 0.5 µm, 8 at 1 µm, 0 at 5 µm, and 0 at 10 µm.

4. **ISO Class 4:**

   - Maximum allowable particles/m³: 10,000 at 0.1 µm, 2,370 at 0.2 µm, 1,020 at 0.3 µm, 352 at 0.5 µm, 83 at 1 µm, 2 at 5 µm, and 0 at 10 µm.

5. **ISO Class 5:**

   - Maximum allowable particles/m³: 100,000 at 0.1 µm, 23,700 at 0.2 µm, 10,200 at 0.3 µm, 3,520 at 0.5 µm, 832 at 1 µm, 29 at 5 µm, and 2 at 10 µm.

The ISO cleanroom classification is determined by the cleanliness level required for the specific manufacturing or research processes taking place within the cleanroom. Lower ISO classes indicate cleaner environments with stricter particle count limits.

## 17.4 Compare Between ISO 14644-1 and Federal Standard 209E.

| Aspect | Federal Standard 209E | ISO 14644-1 |
| --- | --- | --- |
| Terminology | Cleanroom and Controlled Environment | Cleanrooms and Associated Controlled Environments |
| Particle Count Limits (Class 1 to Class 100,000) | Maximum allowable particles per cubic foot of air | Maximum allowable particles per cubic meter of air |
| Particle Size (micrometers) | 0.5, 1.0, 5.0, 10.0 | 0.1, 0.2, 0.3, 0.5, 1.0, 5.0, 10.0, 25.0 |
| Sampling Method | Single-point sampling | Multiple-location sampling |
| Classification Scheme | Integer classification (e.g., Class 100, Class 1,000) | Decimal classification (e.g., ISO 5, ISO 6) |
| Number of Classes | Up to Class 100,000 | Up to ISO Class 9 |
| Cleanroom Monitoring | Air changes per hour (ACH) | Particle count and microbial monitoring |
| Cleanroom Designation | Designated by the number of particles per cubic foot | Designated by the maximum allowable particle counts |

## 17.5 Environmental Conditions Inside semiconductor cleanrooms

The specific temperature, pressure, and humidity conditions inside a semiconductor cleanroom can vary depending on the manufacturing process and the requirements of the semiconductor fabrication facility. However, here are typical ranges for these parameters:

1. **Temperature:**

    Cleanrooms are maintained at controlled temperatures to ensure stability and to prevent variations in manufacturing processes. The temperature inside a semiconductor cleanroom typically falls within the range of 20 to 24 degrees Celsius (68 to 75 degrees Fahrenheit).

2. **Pressure:**

    Cleanrooms are often maintained at a slightly higher pressure than the surrounding areas to prevent contaminants from entering. Positive pressure is commonly used to keep particles from infiltrating the cleanroom. The pressure inside a semiconductor cleanroom is usually in the range of 5 to 30 pascals above atmospheric pressure.

3. **Humidity:**

- Humidity levels are carefully controlled in semiconductor cleanrooms to prevent electrostatic discharge and to maintain the integrity of sensitive semiconductor components. The relative humidity (RH) in a semiconductor cleanroom typically falls within the range of 30% to 70%. Specific processes may have more stringent requirements, and adjustments are made accordingly.

It's important to note that these values can be adjusted based on the specific manufacturing processes, the equipment used, and the materials being processed within the cleanroom. Cleanroom specifications are critical in semiconductor manufacturing to ensure the quality and reliability of the fabricated electronic devices.

# Chapter 18: Cleanroom Environment Experience

## 18.1 HEPA filters

High-Efficiency Particulate Air (HEPA) filters are critical components of cleanroom facilities, playing a vital role in maintaining the cleanliness and integrity of controlled environments. In this section, we explore the importance of HEPA filters, their validation processes, and their contribution to ensuring optimal air quality within cleanroom environments.

### 1. Importance of HEPA Filters:

HEPA filters are designed to capture and remove airborne particles, contaminants, and microorganisms from the air, thereby minimizing the risk of contamination in cleanroom environments. Their high-efficiency filtration capabilities enable HEPA filters to trap particles as small as 0.3 microns with a filtration efficiency of 99.97% or higher, making them highly effective in removing a wide range of airborne pollutants.

The presence of airborne particles, such as dust, microbes, and other contaminants, poses a significant threat to product quality, process integrity, and personnel safety within cleanroom facilities. HEPA filters act as the first line of defense against airborne contamination, ensuring that cleanroom air meets the stringent cleanliness standards required for sensitive manufacturing processes, research activities, and healthcare applications.

### 2. Validation of HEPA Filters:

The performance and integrity of HEPA filters are validated through rigorous testing procedures to ensure compliance with regulatory requirements and industry standards. HEPA filter validation typically involves the following steps:

- **Integrity Testing:** Integrity testing is conducted to verify the effectiveness of HEPA filters in capturing and retaining airborne particles. Common integrity testing methods include aerosol challenge tests, DOP (dioctyl phthalate) or PSL (polystyrene latex) particle penetration tests, and photometer tests. These tests assess factors such as filter efficiency, leakage rates, and pressure differentials across the filter media.

- **Leak Testing:** Leak testing is performed to detect and quantify any leaks or breaches in the seal integrity of HEPA filters. Methods such as the DOP or PSL aerosol leak test and the smoke visualization test are used to identify leaks and ensure that HEPA filters maintain their specified performance characteristics.

- **Installation Qualification (IQ) and Operational Qualification (OQ):** IQ and OQ procedures verify that HEPA filters are installed correctly and operate effectively within cleanroom systems. These qualification tests include verification of filter installation, airflow velocity and uniformity testing, pressure drop measurements, and documentation of compliance with design specifications and performance criteria.

- **Periodic Testing and Monitoring:** Regular testing and monitoring of HEPA filters are essential to ensure ongoing performance and compliance with cleanliness standards. Periodic testing may include airflow velocity measurements, filter integrity tests, and visual inspections to detect any signs of damage, degradation, or contamination.

**3. Maintenance and Replacement:**

Proper maintenance and timely replacement of HEPA filters are essential for ensuring continued effectiveness and reliability in cleanroom operations. Routine maintenance activities include cleaning or replacing pre-filters, monitoring pressure differentials across filters, and scheduling filter replacements based on recommended service intervals or performance degradation indicators.

When replacing HEPA filters, it is essential to use filters that meet or exceed the specifications and performance criteria of the original filters. Improper selection or installation of replacement filters can compromise cleanroom air quality and jeopardize process integrity. Cleanroom managers should work closely with qualified suppliers and manufacturers to ensure the compatibility and suitability of replacement filters for their specific cleanroom applications.

**In summary**, HEPA filters are indispensable components of cleanroom facilities, providing critical protection against airborne contaminants and ensuring the cleanliness and integrity of controlled environments. Through rigorous validation, maintenance, and monitoring processes, cleanroom managers can ensure the effective performance and reliability of HEPA filters, thereby safeguarding product quality, process integrity, and personnel safety within cleanroom environments.

## 18.2 Laminar Flow in Cleanroom Environments

Laminar flow is a fundamental principle of airflow management in cleanroom environments, contributing significantly to the maintenance of high levels of cleanliness and contamination control. In this section, we explore the description of laminar flow, how it works, and its critical importance in ensuring the integrity and functionality of cleanroom environments.

## 1. Description of Laminar Flow:

Laminar flow refers to the controlled and uniform movement of air in parallel layers or streams, with minimal turbulence or mixing between adjacent layers. In a laminar flow system, air moves smoothly and steadily in a single direction, typically from a high-efficiency particulate air (HEPA) filter or air supply outlet towards a return or exhaust vent. This unidirectional flow pattern ensures that airborne particles, contaminants, and microorganisms are carried away from critical work areas and personnel, maintaining a clean and controlled environment.

## 2. How Laminar Flow Works:

Laminar flow is achieved through the careful design and configuration of airflow systems within cleanroom facilities. Key components of a laminar flow system include:

- **HEPA Filters:** High-efficiency particulate air (HEPA) filters serve as the primary source of clean air in cleanroom environments. These filters remove airborne particles and contaminants from incoming air, ensuring that the air entering the cleanroom is of the highest possible quality and cleanliness.

- **Airflow Direction:** In a laminar flow system, air flows in a controlled and uniform direction, typically from the ceiling to the floor or from one end of the cleanroom to the other. This unidirectional airflow pattern prevents the recirculation of contaminated air and minimizes the risk of cross-contamination between different areas within the cleanroom.

- **Airflow Velocity and Uniformity:** Laminar flow systems maintain consistent airflow velocities and uniformity throughout the cleanroom environment. Airflow velocity is carefully controlled to ensure that particles and contaminants are effectively carried away from critical work areas and personnel, while maintaining comfortable working conditions.

- **Ceiling-mounted Supply Outlets:** HEPA-filtered air is supplied into the cleanroom environment through ceiling-mounted supply outlets or vents. These outlets release clean air in a downward or horizontal direction, creating a continuous flow of clean air that sweeps across the workspace and carries away contaminants.

- **Return or Exhaust Vents:** Used air, along with any contaminants it may contain, is drawn out of the cleanroom environment through return or exhaust vents located near the floor or at designated exit points. This ensures that contaminated air is effectively removed from the cleanroom, maintaining the integrity of the airflow pattern and cleanliness levels.

**3. Critical Importance of Laminar Flow:**

Laminar flow is of critical importance in cleanroom environments for several reasons:

- **Contamination Control:** Laminar flow systems minimize the risk of airborne contamination by continuously supplying clean, filtered air to critical work areas while removing contaminated air away from personnel and sensitive processes. This helps maintain cleanliness levels within the cleanroom and prevents the accumulation of particles, microbes, and other contaminants that could compromise product quality or safety.

- **Process Integrity:** The uniform and controlled airflow provided by laminar flow systems ensures the consistency and reliability of manufacturing processes conducted within cleanroom environments. By minimizing airflow disturbances and fluctuations, laminar flow systems help maintain stable environmental conditions essential for sensitive processes such as semiconductor fabrication, pharmaceutical manufacturing, and biotechnology research.

- **Personnel Protection:** Laminar flow systems protect cleanroom personnel from exposure to airborne contaminants and hazardous substances by directing clean air towards work areas and away from personnel. This helps safeguard the health and safety of cleanroom workers and minimizes the risk of occupational exposure to potentially harmful substances.

- **Regulatory Compliance:** Laminar flow systems are designed to meet regulatory requirements and industry standards for cleanroom cleanliness and contamination control. Compliance with these standards is essential for obtaining regulatory approvals, maintaining product quality, and ensuring customer satisfaction in industries such as pharmaceuticals, healthcare, and electronics manufacturing.

**In summary**, laminar flow is a critical component of cleanroom environments, providing controlled and uniform airflow to maintain high levels of cleanliness, contamination control, and process integrity. By effectively managing airflow patterns and directing clean air towards critical work areas, laminar flow systems contribute to the safety, reliability, and regulatory compliance of cleanroom operations across a wide range of industries and applications.

## 18.3 HVAC System in Cleanroom Environments

Heating, Ventilation, and Air Conditioning (HVAC) systems are integral components of cleanroom environments, playing a crucial role in maintaining optimal environmental conditions, cleanliness levels, and air quality. In this detailed section, we explore the key components, design considerations, operation principles, and critical importance of HVAC systems in cleanroom facilities.

**1. Key Components of HVAC Systems:**

HVAC systems in cleanroom environments typically consist of the following key components:

**Air Handling Units (AHUs):** AHUs are central components of HVAC systems responsible for conditioning and distributing air within cleanroom facilities. These units include components such as fans, filters, coils, and dampers, which work together to regulate airflow, temperature, humidity, and cleanliness levels.

**Filters:** High-efficiency particulate air (HEPA) filters and/or Ultra-Low Penetration Air (ULPA) filters are used in HVAC systems to remove airborne particles, contaminants, and microorganisms from incoming air. These filters are typically installed upstream of AHUs to ensure that the air entering the cleanroom is of the highest possible quality and cleanliness.

**Ductwork:** Ductwork systems distribute conditioned air from AHUs to cleanroom spaces, ensuring uniform airflow distribution and maintaining consistent environmental conditions throughout the facility. Ductwork may include supply ducts, return ducts, and exhaust ducts, as well as dampers and louvers for airflow control and modulation.

**Air Distribution Devices:** Air distribution devices, such as diffusers, grilles, registers, and terminal filters, are installed at strategic locations within cleanroom spaces to distribute conditioned air effectively and efficiently. These devices help maintain laminar airflow patterns, control air velocity, and minimize turbulence, ensuring optimal cleanliness levels and environmental stability.

**Controls and Monitoring Systems:** HVAC systems are equipped with sophisticated control and monitoring systems to regulate and monitor environmental parameters, such as temperature, humidity, pressure differentials, and airflow rates. These systems employ sensors, actuators, controllers, and human-machine interfaces (HMIs) to maintain precise control over HVAC operations and ensure compliance with cleanliness standards and regulatory requirements.

## 2. Design Considerations for HVAC Systems:

Designing HVAC systems for cleanroom environments requires careful consideration of various factors to ensure optimal performance, reliability, and compliance with cleanliness standards. Key design considerations include:

- **Cleanliness Requirements:** HVAC systems must be designed to meet the cleanliness requirements of the cleanroom classification and the specific needs of the application. This includes selecting appropriate filtration technology, airflow patterns, and control strategies to achieve and maintain the desired cleanliness levels.

- **Airflow Patterns:** The design of airflow patterns, such as laminar flow or turbulent flow, depends on the requirements of the cleanroom application and the sensitivity of the processes conducted within the facility. Laminar airflow patterns are commonly used in cleanrooms to minimize the risk of cross-contamination and ensure uniform distribution of clean air.

- **Temperature and Humidity Control:** HVAC systems must maintain precise control over temperature and humidity levels to create a stable and comfortable working environment within the cleanroom. Temperature and humidity control are critical for ensuring product quality, process integrity, and personnel comfort in cleanroom environments.

- **Pressure Differentials:** Positive or negative pressure differentials may be required between cleanroom spaces or between cleanroom areas and adjacent non-clean areas to prevent the ingress of contaminants and maintain cleanliness levels. HVAC systems must be designed to maintain and control pressure differentials effectively to achieve the desired environmental conditions.

- **Energy Efficiency:** Energy-efficient design principles, such as variable-speed drives, energy recovery systems, and optimized control strategies, should be incorporated into HVAC systems to minimize energy consumption and operating costs while maximizing performance and sustainability.

## 3. Operation Principles of HVAC Systems:

HVAC systems operate based on the principles of air conditioning, filtration, and air distribution to maintain optimal environmental conditions within cleanroom facilities. The operation of HVAC systems involves the following principles:

- **Air Conditioning:** HVAC systems condition incoming air by cooling, heating, dehumidifying, or humidifying it to achieve the desired temperature and humidity levels within the cleanroom. This ensures that cleanroom environments remain within the specified operating range required for the processes conducted within the facility.

- **Filtration:** HVAC systems use HEPA filters or ULPA filters to remove airborne particles, contaminants, and microorganisms from incoming air, ensuring that the air entering the cleanroom is clean and free of contaminants. Filtration plays a critical role in maintaining cleanliness levels and preventing contamination of sensitive processes and products.

- **Air Distribution:** HVAC systems distribute conditioned air throughout cleanroom spaces using ductwork, air distribution devices, and airflow control mechanisms. Uniform airflow distribution helps maintain laminar airflow patterns, minimize turbulence, and ensure consistent environmental conditions across the cleanroom facility.

- **Monitoring and Control:** HVAC systems are equipped with sensors, controllers, and monitoring devices to measure environmental parameters such as temperature, humidity, pressure, and airflow rates. These systems continuously monitor environmental conditions and adjust HVAC operations to maintain optimal conditions within the cleanroom.

**4. Critical Importance of HVAC Systems in Cleanroom Environments:**

HVAC systems play a critical role in cleanroom environments for the following reasons:

- **Contamination Control:** HVAC systems are essential for controlling airborne contaminants and maintaining cleanliness levels within cleanroom environments. By filtering and conditioning incoming air, HVAC systems help minimize the risk of contamination and ensure the integrity of sensitive processes and products conducted within the facility.

- **Environmental Stability:** HVAC systems create and maintain stable environmental conditions, including temperature, humidity, and airflow, required for the successful operation of cleanroom facilities. Consistent environmental conditions are essential for ensuring process integrity, product quality, and regulatory compliance in industries such as pharmaceuticals, electronics, and biotechnology.

- **Personnel Comfort and Safety:** HVAC systems provide a comfortable and safe working environment for cleanroom personnel by regulating temperature, humidity, and airflow levels. Proper environmental control helps prevent heat stress, humidity-related discomfort, and airborne exposure to contaminants, ensuring the well-being and productivity of cleanroom workers.

- **Regulatory Compliance:** HVAC systems must comply with regulatory requirements, cleanliness standards, and industry guidelines governing cleanroom operations. Compliance with these regulations is essential for obtaining regulatory approvals, maintaining product quality, and ensuring the safety of personnel and the environment.

**In summary,** HVAC systems are essential components of cleanroom environments, providing critical functions such as conditioning, filtration, air distribution, and environmental control. By effectively managing environmental conditions and controlling airborne contaminants, HVAC systems contribute to the cleanliness, integrity, and functionality of cleanroom facilities across diverse industries and applications.

## 18.4 Deionized Water System in Cleanroom Environments

Deionized water, often abbreviated as DI water, plays a critical role in various processes within cleanroom environments, ranging from equipment cleaning to chemical formulations. In this section, we explore the characteristics of DI water, its usage, importance, generation process, validation, and maintenance within cleanroom facilities.

**1. Characteristics of Deionized Water:**

Deionized water is purified water that has undergone a process to remove ions and impurities, resulting in a high level of purity and electrical resistance. Some key characteristics of deionized water include:

- **Purity:** Deionized water is virtually free from dissolved minerals, salts, and other contaminants, making it suitable for sensitive applications where even trace impurities could cause issues.

- **Electrical Resistance:** DI water exhibits high electrical resistance due to the absence of ions. This property is essential in applications where conductivity must be minimized, such as in electronics manufacturing or laboratory experiments.

**2. Usage and Importance of Deionized Water:**

Deionized water finds widespread usage in cleanroom environments for various purposes, including:

- **Equipment Cleaning:** DI water is commonly used for cleaning equipment, tools, and surfaces within cleanrooms to ensure cleanliness and prevent contamination of sensitive processes and products.

- **Chemical Formulations:** DI water serves as a solvent or diluent in chemical formulations, analytical procedures, and experimental setups, where the absence of impurities is crucial for accurate results and reproducibility.

- **Rinsing and Final Cleaning:** DI water is used for rinsing and final cleaning steps in manufacturing processes, such as semiconductor fabrication, pharmaceutical production, and medical device manufacturing, to remove residual chemicals or particles and ensure product purity.

- **Analytical Instrumentation:** DI water is used as a solvent or blank in analytical instrumentation, such as spectrophotometers, chromatography systems, and titration setups, where purity and consistency are essential for accurate measurements.

## 3. Generation of Deionized Water using Reverse Osmosis Systems:

Deionized water is typically generated using reverse osmosis (RO) systems, which employ semi-permeable membranes to remove ions, dissolved solids, and contaminants from water. The process of generating deionized water using reverse osmosis systems involves the following steps:

- **Pre-Treatment:** Incoming water is pre-treated to remove suspended solids, organic matter, chlorine, and other impurities that could foul or damage the RO membranes. Pre-treatment methods may include sediment filtration, carbon filtration, and chemical dosing.

- **Reverse Osmosis:** Pre-treated water is pressurized and forced through semi-permeable RO membranes, which selectively allow water molecules to pass while rejecting ions, dissolved solids, and contaminants. The purified water, known as permeate, is collected, while the rejected impurities are discharged as concentrate or reject.

- **Deionization:** The permeate from the RO system may undergo further treatment using ion exchange resins to remove any remaining ions and achieve the desired level of purity. This process, known as deionization or polishing, ensures that the resulting water meets the stringent purity requirements of cleanroom applications.

## 4. Importance of DI Water Validation and Maintenance:

Validation and maintenance of DI water systems are essential to ensure the reliability, consistency, and quality of the generated water. Key aspects of DI water validation and maintenance include:

- **Performance Verification:** Regular testing and validation of DI water quality, conductivity, and purity are necessary to verify that the system is operating within specified parameters and meeting cleanliness standards for cleanroom applications.

- **Calibration and Monitoring:** Calibration and monitoring of instrumentation, sensors, and control systems are essential to ensure accurate measurement and control of water quality parameters such as conductivity, pH, and total dissolved solids (TDS).

- **Consumables Replacement:** Periodic replacement of consumables such as RO membranes, ion exchange resins, filters, and cartridges is necessary to maintain system performance and prevent degradation of water quality over time.

- **Cleaning and Sanitization:** Routine cleaning and sanitization of DI water system components, including RO membranes, piping, tanks, and filters, are essential to prevent fouling, biofilm formation, and microbial contamination that could compromise water quality and system performance.

**In summary**, deionized water systems play a critical role in cleanroom environments, providing high-purity water for various applications where cleanliness, purity, and electrical resistance are paramount. By understanding the characteristics, usage, generation process, validation, and maintenance requirements of DI water systems, cleanroom operators can ensure the consistent supply of high-quality water essential for successful cleanroom operations and manufacturing processes.

## 18.5 Importance of Liquid Nitrogen Tanks in Cleanroom Operations

Liquid nitrogen (LN2) tanks play a crucial role in maintaining optimal conditions and supporting various processes within cleanroom environments. From providing pneumatic gas sources to facilitating equipment cleaning, LN2 tanks are indispensable assets. In this detailed section, we explore the multifaceted importance of LN2 tanks, their usage, maintenance, and associated considerations within cleanroom facilities.

**1. Pneumatic Gas Source:**

LN2 tanks serve as pneumatic gas sources for gas cabinets, supplying nitrogen gas for a wide range of applications within cleanroom environments. Nitrogen gas is commonly used for purging, blanketing, and pressurizing processes, ensuring an inert atmosphere and preventing contamination in sensitive manufacturing processes.

## 2. Venting Gas Cabinets:

LN2 is utilized for venting gas cabinets, purging potentially hazardous or reactive gases from the system to maintain safety and prevent cross-contamination between different processes or tool chambers within the cleanroom.

## 3. Alternative to Clean Dry Air (CDA):

LN2 can serve as an alternative to Clean Dry Air (CDA) in cleanroom tools and equipment where inert gas is required. Its inert properties make it suitable for applications where oxygen or moisture contamination must be minimized to maintain product integrity.

## 4. Cleaning Applications:

LN2 is used in cleaning processes within cleanrooms, providing a rapid and effective method for removing contaminants, residues, and particulates from equipment, surfaces, and components. Its low temperature and high vaporization rate enable efficient cleaning without leaving behind any residues or moisture.

## 5. Degree of Purity:

The purity of LN2 is critical for ensuring the quality and integrity of processes within cleanroom environments. LN2 purity levels typically range from 99.9% to 99.999%, depending on the application requirements and quality standards.

## 6. Dewar Filling Kits:

Dewar filling kits are used to transfer LN2 from bulk storage tanks to smaller dewars or containers for convenient handling and distribution within the cleanroom. These kits include hoses, connectors, valves, and safety features to ensure safe and efficient transfer of LN2.

## 7. Tank Filling Rate and Monitoring:

LN2 tanks are filled at a controlled rate to prevent overfilling and ensure safety. Filling companies utilize automatic monitoring systems to track LN2 levels and anticipate when refilling will be required based on usage patterns and demand.

## 8. Tank Maintenance Responsibility:

Maintenance of LN2 tanks is the responsibility of cleanroom personnel, ensuring that tanks are properly inspected, cleaned, and serviced at regular intervals to prevent leaks, contamination, or other safety hazards. Routine maintenance may include visual inspections, pressure testing, and valve maintenance.

## 9. Calculating Needed Volume (Net Capacity) of the Tank:

The needed volume or net capacity of an LN2 tank is calculated based on factors such as the rate of LN2 consumption, frequency of usage, and duration of operation. Cleanroom operators can estimate the required volume by analyzing historical usage data and projected future demand.

**10. Vaporizer and Working Pressure:**

LN2 tanks may be equipped with vaporizers to convert liquid nitrogen into gaseous form for distribution to gas cabinets or equipment. The working pressure of the vaporizer is calculated based on factors such as flow rate, temperature, and desired gas pressure, ensuring optimal performance and safety.

**11. Continuous Flow Rate and Step Down Non-Return Circuit:**

The continuous flow rate of LN2 from the tank to gas cabinets or equipment is regulated using a step-down non-return circuit, which controls the release of gas at a consistent pressure and flow rate. This circuit ensures stable gas delivery and prevents backflow or pressure fluctuations.

**In summary**, liquid nitrogen tanks are indispensable assets in cleanroom operations, providing a reliable and versatile source of inert gas for various applications. By understanding their usage, maintenance requirements, and associated considerations, cleanroom personnel can optimize the performance and safety of LN2 systems, ensuring seamless operations and process integrity within cleanroom environments.

## 18.6 Measurement of Contamination Levels in Cleanroom Environments

Measuring contamination levels in a cleanroom environment is essential for ensuring compliance with cleanliness standards, maintaining product quality, and safeguarding sensitive processes. In this section, we explore the methods used to measure contamination levels, the frequency of monitoring, and the role of the cleanroom manager in response to high contamination levels.

**1. Methods for Measuring Contamination Levels:**

Several methods are employed to measure contamination levels in cleanroom environments, including:

- **Particle Counting:** Particle counters are used to measure the concentration and size distribution of airborne particles in cleanroom air. These devices utilize light scattering or light extinction techniques to detect and quantify particles ranging in size from sub-micron to larger sizes.

- **Microbiological Monitoring:** Microbiological monitoring involves sampling and culturing air, surfaces, or liquids within the cleanroom environment to detect the presence of microorganisms such as bacteria, fungi, and yeast. Microbial enumeration techniques, such as agar plate counts and membrane filtration, are used to assess microbial contamination levels.

- **Surface Contamination Monitoring:** Surface sampling techniques, such as swabbing or contact plates, are employed to assess the cleanliness of surfaces within the cleanroom environment. These methods detect the presence of viable or non-viable contaminants on surfaces, equipment, or tools.

- **Chemical Contamination Monitoring:** Chemical contamination levels in cleanroom environments can be measured using analytical techniques such as gas chromatography, mass spectrometry, or ion chromatography. These methods identify and quantify chemical contaminants, residues, or impurities present in the air, water, or surfaces.

## 2. Frequency of Monitoring:

The frequency of contamination monitoring in cleanroom environments depends on factors such as cleanliness requirements, process sensitivity, industry standards, and regulatory guidelines. Common practices for monitoring contamination levels include:

- **Continuous Monitoring:** Some cleanrooms are equipped with continuous monitoring systems that provide real-time data on contamination levels, allowing for immediate detection of deviations from cleanliness standards.

- **Periodic Monitoring:** Periodic contamination monitoring is conducted at regular intervals, such as daily, weekly, or monthly, to assess long-term trends, identify potential sources of contamination, and verify the effectiveness of cleaning and maintenance procedures.

- **Event-Based Monitoring:** Contamination monitoring may be triggered by specific events or activities within the cleanroom environment, such as equipment maintenance, process changes, or environmental disturbances. These events may require immediate monitoring to assess their impact on contamination levels.

## 3. Cleanroom Manager's Role in Response to High Contamination Levels:

When high contamination levels are detected in a cleanroom environment, the cleanroom manager plays a critical role in implementing corrective actions and mitigating potential risks. Responsibilities may include:

- **Immediate Response:** The cleanroom manager must respond promptly to high contamination levels, assessing the situation, identifying the root cause, and implementing corrective measures to reduce contamination levels and prevent further escalation.

- **Root Cause Analysis:** Conducting a thorough investigation to determine the underlying causes of high contamination levels, such as equipment malfunction, process deviations, or human error. Identifying and addressing root causes is essential for preventing recurrence and improving overall cleanliness.

- **Documentation and Reporting:** Documenting contamination incidents, corrective actions taken, and follow-up measures for future reference and regulatory compliance. Reporting contamination events to relevant stakeholders, management, or regulatory authorities as necessary.

- **Review and Improvement:** Reviewing contamination monitoring data, trends, and historical records to identify areas for improvement in cleanroom operations, maintenance practices, and contamination control strategies. Implementing corrective actions and preventive measures to enhance cleanliness levels and mitigate contamination risks.

**In summary**, measuring contamination levels in cleanroom environments is essential for ensuring cleanliness standards, product quality, and process integrity. By employing appropriate monitoring methods, frequency, and response protocols, cleanroom managers can effectively manage contamination risks and maintain the required level of cleanliness within the facility.

## 18.7 Measuring Oxygen Levels in a Cleanroom Environment

Measuring oxygen levels in a cleanroom environment is crucial for ensuring the safety and comfort of personnel, as well as maintaining optimal conditions for processes and equipment. In this section, we explore the units of measurement, tools used, importance of measuring oxygen levels, and the role of the cleanroom manager in addressing low oxygen level situations.

**1. Units of Measurement:**

Oxygen levels in cleanroom environments are typically measured in terms of partial pressure or concentration, using units such as:

- **Partial Pressure:** Oxygen partial pressure is measured in units of millimeters of mercury (mmHg) or kilopascals (kPa), representing the pressure exerted by oxygen molecules in the air.

- **Concentration:** Oxygen concentration is measured as a percentage of the total gas mixture, often expressed as volume percent (% vol) or parts per million (ppm) by volume.

## 2. Tools Used for Measurement:

Several tools and instruments are available for measuring oxygen levels in cleanroom environments, including:

- **Oxygen Analyzers:** Portable or fixed oxygen analyzers are used to directly measure oxygen concentration in the air. These analyzers utilize electrochemical sensors, paramagnetic sensors, or infrared sensors to detect and quantify oxygen levels accurately.

- **Gas Detectors:** Multi-gas detectors equipped with oxygen sensors can monitor oxygen levels along with other gases such as combustible gases, toxic gases, and volatile organic compounds (VOCs). These detectors provide real-time readings and alarms for oxygen depletion situations.

## 3. Importance of Measuring Oxygen Levels:

Measuring oxygen levels in a cleanroom environment is important for several reasons:

- **Safety:** Monitoring oxygen levels ensures the safety of personnel working in the cleanroom by alerting them to potential oxygen-deficient atmospheres, which can pose risks of asphyxiation or respiratory distress.

- **Process Control:** Oxygen levels can affect the performance and stability of processes and equipment within the cleanroom. Monitoring oxygen levels helps maintain optimal conditions for sensitive processes, such as semiconductor fabrication, pharmaceutical manufacturing, and biological research.

- **Comfort and Productivity:** Adequate oxygen levels contribute to the comfort and well-being of cleanroom personnel, promoting productivity, alertness, and cognitive function. Monitoring and maintaining oxygen levels within acceptable ranges support a healthy and productive working environment.

## 4. Role of the Cleanroom Manager:

In the event of low oxygen levels in a cleanroom environment, the cleanroom manager plays a critical role in addressing the situation and implementing corrective measures. Responsibilities may include:

- **Immediate Response:** The cleanroom manager must respond promptly to low oxygen level alarms or indications, assessing the situation and ensuring the safety of personnel within the cleanroom.

- **Ventilation Adjustment:** Adjusting ventilation systems, including HVAC systems and air exchange rates, to increase the supply of fresh air and oxygen within the cleanroom. Increasing ventilation rates can help restore oxygen levels to safe and comfortable levels.

- **Emergency Procedures:** Implementing emergency procedures, such as evacuating personnel from the affected area, providing supplemental oxygen, and contacting emergency response teams if necessary.

- **Investigation and Prevention:** Conducting a thorough investigation to determine the cause of low oxygen levels, such as ventilation system malfunction, air supply issues, or chemical reactions. Implementing corrective actions and preventive measures to prevent recurrence and improve oxygen level monitoring and control.

**In summary**, measuring oxygen levels in a cleanroom environment is essential for ensuring safety, process control, and personnel well-being. By employing appropriate measurement tools, monitoring techniques, and response protocols, cleanroom managers can effectively manage oxygen levels and address low oxygen level situations to maintain a safe and productive working environment within the facility.

## 18.8 Importance of Uninterruptible Power Supply (UPS) in Cleanroom Environments

Uninterruptible Power Supply (UPS) systems are critical components in cleanroom environments, providing backup power and ensuring uninterrupted operation of essential equipment and processes. In this section, we delve into the importance of UPS systems, load balancing strategies, and prioritization of loads in cleanroom settings.

**1. Importance of UPS Systems:**

UPS systems play a vital role in maintaining continuity and reliability in cleanroom operations for the following reasons:

- **Equipment Protection:** UPS systems protect sensitive equipment and processes from power disturbances, including voltage sags, surges, spikes, and momentary outages. By providing a stable and clean power supply, UPS systems prevent equipment damage, data loss, and process interruptions.

- **Process Integrity:** Cleanroom processes, such as semiconductor fabrication, pharmaceutical manufacturing, and precision instrumentation, require stable and uninterrupted power to maintain product quality, consistency, and yield. UPS systems ensure process integrity by eliminating the risk of power-related disruptions or anomalies.

- **Data Preservation:** Cleanroom environments often house critical data acquisition systems, control systems, and monitoring equipment. UPS systems prevent data loss and corruption by providing backup power during unexpected power outages, allowing for seamless data acquisition and recording.

- **Compliance Requirements:** Regulatory standards and industry guidelines mandate the use of UPS systems in cleanroom environments to ensure continuity of critical operations, compliance with safety regulations, and adherence to quality standards. UPS systems help cleanroom facilities meet regulatory requirements and maintain certification.

**2. Load Balancing Strategies:**

Load balancing is essential for optimizing UPS performance and maximizing system reliability in cleanroom environments. Key strategies for load balancing include:

- **Equal Distribution:**

  Distributing electrical loads evenly across UPS units or phases to prevent overloading of individual units and ensure efficient utilization of available capacity. Equal distribution minimizes the risk of overloading, overheating, and premature equipment failure.

- **Redundancy Planning:**

  Implementing redundant UPS configurations, such as N+1 or N+2 redundancy, to provide backup power redundancy and fault tolerance. Redundancy planning ensures continuous operation even in the event of UPS unit failure or maintenance downtime.

- **Monitoring and Management:**

  Utilizing UPS monitoring and management software to monitor load levels, battery health, and system status in real-time. Monitoring tools provide valuable insights into UPS performance, allowing for proactive load balancing and capacity planning.

**3. Prioritization of Loads:**

Prioritizing loads in cleanroom environments ensures that critical equipment and processes receive priority power supply during UPS operation. Key considerations for load prioritization include:

- **Criticality of Equipment:**

  Identifying critical equipment and processes that are essential for cleanroom operations, safety, and compliance. Critical loads, such as HVAC systems, process equipment, and emergency lighting, should receive priority power supply from the UPS.

- **Emergency Lighting:**

  Prioritizing emergency lighting loads to ensure visibility and safety in the event of a power outage or evacuation. Emergency lighting systems should be connected to UPS units with dedicated battery backup to maintain illumination during power disturbances.

- **Sequential Startup:**

  Implementing sequential startup sequences for non-essential equipment and lighting loads to minimize initial power demand and reduce the risk of overload during UPS switchover. Sequential startup ensures a smooth transition to backup power without overloading the UPS system.

**In summary**, UPS systems are essential for ensuring continuity, reliability, and compliance in cleanroom environments. By implementing load balancing strategies, prioritizing critical loads, and maintaining redundancy, cleanroom operators can optimize UPS performance and mitigate the impact of power disturbances on sensitive equipment and processes within the facility.

## 18.9 Fire Fighting Systems in Cleanroom Environments

Fire fighting systems in cleanroom environments are designed to mitigate the risk of fire and minimize potential damage to sensitive equipment, processes, and personnel while maintaining the cleanliness and integrity of the controlled environment. In this section, we explore the nature of fire fighting systems tailored for cleanroom settings, focusing on their design principles, components, and operation.

**1. Design Principles:**

Fire fighting systems in cleanrooms are designed with several key principles in mind to ensure effectiveness, reliability, and compatibility with cleanroom requirements:

- **Minimal Contamination:**

Fire suppression agents and system components are selected to minimize the risk of contamination to cleanroom environments. Non-corrosive, non-residue-forming agents are preferred to maintain cleanliness and prevent damage to sensitive equipment and processes.

- **Rapid Response:**

Cleanroom fire fighting systems are designed for rapid detection and suppression of fires to minimize downtime and prevent escalation. Early detection methods, such as smoke detectors, heat detectors, or flame detectors, trigger immediate activation of suppression systems for swift intervention.

- **Selective Suppression:**

Cleanroom fire fighting systems employ suppression agents tailored to the specific hazards and requirements of cleanroom environments. Clean agents, such as inert gases (e.g., nitrogen, argon) or clean chemical agents (e.g., FM-200, Novec 1230), are used to extinguish fires without leaving residues or contaminating the cleanroom environment.

## 2. Components of Fire Fighting Systems:

Fire fighting systems in cleanrooms consist of several essential components designed to detect, suppress, and control fires effectively:

- **Detection Systems:**

Smoke detectors, heat detectors, or flame detectors are installed throughout the cleanroom to detect signs of fire and trigger alarm notifications. Early detection systems enable rapid response and intervention to prevent fire spread.

- **Suppression Systems:**

Cleanroom fire suppression systems utilize clean agents, such as inert gases or clean chemical agents, delivered through strategically placed nozzles or sprinklers. These systems extinguish fires quickly and efficiently without leaving residues or damaging equipment.

- **Control Panels:**

Fire control panels serve as the central hub for monitoring and controlling fire suppression systems. They provide real-time status updates, alarm notifications, and remote activation capabilities for rapid response to fire emergencies.

- **Backup Power:**

  Fire fighting systems may be equipped with backup power sources, such as uninterruptible power supply (UPS) units or emergency generators, to ensure continuous operation during power outages or emergencies.

## 3. Operation and Maintenance:

The operation and maintenance of fire fighting systems in cleanrooms are critical for ensuring reliability, compliance, and readiness to respond to fire emergencies:

- **Regular Inspection:**

  Fire suppression systems should be inspected regularly by qualified technicians to verify proper operation, integrity of components, and compliance with regulatory requirements and industry standards.

- **Testing and Certification:**

  Periodic testing and certification of fire fighting systems are essential to validate performance, functionality, and effectiveness in suppressing fires. Testing procedures may include functional tests, flow tests, and agent concentration tests to verify system integrity and performance.

- **Training and Emergency Response:**

  Cleanroom personnel should receive training on fire safety procedures, evacuation routes, and operation of fire fighting equipment. Regular drills and exercises help ensure that personnel are prepared to respond effectively to fire emergencies and minimize risks to life and property.

## 4. Integration with Cleanroom Design:

Fire fighting systems are integrated into the overall design and layout of cleanroom facilities to ensure compatibility with cleanliness requirements, workflow patterns, and safety considerations. System components, such as detectors, nozzles, and control panels, are strategically positioned to provide comprehensive coverage and minimize interference with cleanroom operations.

**In summary**, fire fighting systems in cleanroom environments are designed and implemented with careful consideration of cleanliness, effectiveness, and safety. By employing advanced detection technologies, clean agents, and proactive maintenance practices, cleanroom operators can mitigate the risk of fire and safeguard critical assets and personnel within the controlled environment.

## 18.10 Importance of Air Showers in Cleanroom Entrances

Air showers serve as vital components of cleanroom entrances, playing a crucial role in maintaining the cleanliness and integrity of the controlled environment. In this section, we delve into the importance of air showers, their functionality, and the benefits they offer to cleanroom operations.

**1. Contamination Control:**

Air showers are designed to control contamination by removing particulate matter, dust, and other contaminants from personnel and materials before entering the cleanroom environment. By subjecting individuals and objects to high-velocity, filtered air streams, air showers effectively remove surface contaminants, preventing their introduction into the cleanroom.

**2. Personnel Cleanliness:**

Ensuring the cleanliness of personnel entering the cleanroom is essential for preventing contamination of sensitive processes and products. Air showers help maintain personnel cleanliness by removing loose particles, lint, hair, and skin flakes from clothing and exposed surfaces, minimizing the risk of shedding contaminants within the cleanroom.

**3. Product Protection:**

In industries such as semiconductor manufacturing, pharmaceutical production, and biotechnology, even tiny particles or microbes can compromise product quality and integrity. Air showers provide an additional layer of protection by reducing the likelihood of product contamination caused by personnel or materials entering the cleanroom environment.

**4. Equipment and Surface Cleanliness:**

In addition to personnel, air showers can also be used to clean equipment, tools, and materials before they are brought into the cleanroom. By removing surface contaminants from these items, air showers help maintain the cleanliness of cleanroom surfaces and minimize the risk of cross-contamination between different areas or processes.

**5. Enhanced Airflow Management:**

Air showers contribute to the overall airflow management and contamination control strategy within the cleanroom. By directing high-velocity, filtered air streams at individuals and objects from all directions, air showers create a barrier that helps contain contaminants and prevent their migration into the cleanroom environment.

**6. Compliance with Regulatory Standards:**

Many industries, such as pharmaceuticals, biotechnology, and microelectronics, are subject to stringent regulatory requirements governing cleanliness and contamination control. Air showers are often a mandatory component of cleanroom design and operation, helping facilities comply with regulatory standards and maintain certification.

**7. Improved Cleanroom Performance:**

By reducing the introduction of contaminants into the cleanroom environment, air showers contribute to improved cleanroom performance, stability, and reliability. Cleaner environments lead to fewer process deviations, lower reject rates, and higher product yields, ultimately enhancing overall operational efficiency and profitability.

**8. Employee Health and Safety:**

Air showers also contribute to employee health and safety by minimizing exposure to potentially harmful contaminants and allergens. By removing particles and pollutants from clothing and skin, air showers help create a healthier and safer working environment for cleanroom personnel.

**In summary**, air showers play a critical role in contamination control and cleanliness maintenance within cleanroom environments. By effectively removing surface contaminants from personnel, materials, and equipment before they enter the cleanroom, air showers help safeguard product quality, enhance operational performance, and ensure compliance with regulatory standards.

## 18.11 Importance of Yellow Light in Lithography Rooms within Cleanrooms

Yellow light plays a pivotal role in lithography rooms housed within cleanroom environments, offering several crucial benefits that contribute to the success of semiconductor fabrication processes. In this section, we delve into the significance of yellow light and its various applications within lithography rooms.

**1. Photoresist Sensitivity:**

Yellow light is used in lithography rooms because it falls outside the spectrum of wavelengths that can trigger the photosensitive reactions in photoresist materials. Photoresists used in lithography processes are sensitive to specific wavelengths of light, typically in the ultraviolet (UV) range. Yellow light, with its longer wavelength, does not affect the photoresist, allowing operators to work safely without risking premature exposure or degradation of the photoresist material.

## 2. Operator Safety:

Yellow light enhances operator safety by providing a visible illumination source that minimizes the risk of eye strain and fatigue during extended periods of work in low-light conditions. Unlike bright white light or UV light, which can cause discomfort and strain when viewed for prolonged periods, yellow light offers a softer, more comfortable illumination that is conducive to focused and efficient work in lithography rooms.

## 3. Contrast Enhancement:

Yellow light enhances contrast on photomasks and other lithography-related materials, making it easier for operators to visualize fine details and patterns. The warm-toned illumination provided by yellow light creates sharper outlines and improves the clarity of features on photomasks, facilitating precise alignment and pattern transfer processes during lithography.

## 4. Process Consistency:

Consistency is critical in lithography processes to ensure uniformity and reproducibility across multiple fabrication runs. Yellow light provides stable and consistent illumination that minimizes variations in exposure conditions, helping to maintain process repeatability and yield consistency. By reducing the impact of ambient lighting fluctuations, yellow light contributes to tighter process control and improved manufacturing outcomes.

## 5. Contamination Control:

Yellow light contributes to contamination control efforts within lithography rooms by minimizing the risk of particulate contamination associated with frequent switching between light sources. Because yellow light remains on continuously throughout lithography operations, there is no need to switch between light sources, reducing the potential for introducing contaminants into the cleanroom environment.

## 6. Regulatory Compliance:

Many semiconductor fabrication facilities are subject to regulatory requirements governing safety, cleanliness, and process control. The use of yellow light in lithography rooms helps facilities comply with regulatory standards by providing a safe, reliable, and effective illumination solution that meets the specific needs of lithography processes while minimizing potential risks to personnel and products.

In summary, yellow light plays a vital role in lithography rooms within cleanroom environments, offering numerous benefits that contribute to the success and efficiency of semiconductor fabrication processes. By providing a safe, consistent, and contrast-enhancing illumination source, yellow light helps operators achieve precise alignment, improve process consistency, and maintain compliance with regulatory standards, ultimately supporting the production of high-quality semiconductor devices.

# Chapter 19: Understanding Vacuum Technology: Measurement Units, Gauges, and Pumps

## 19.1 Introduction to Vacuum Technology

Vacuum technology is indispensable in a multitude of industries and applications, serving as the backbone for processes ranging from semiconductor manufacturing to space exploration. Understanding the principles of vacuum technology is paramount for achieving optimal performance and reliability in vacuum systems.

### 19.1.1 Importance of Vacuum Technology

**Cross-Industry Significance: Highlighting the pervasive influence of vacuum technology across diverse sectors, including:**

- Semiconductor Industry: Critical for the production of microchips and electronic components.

- Pharmaceutical Manufacturing: Essential for the production of medicines and vaccines under sterile conditions.

- Aerospace Engineering: Facilitating the creation of low-pressure environments for space simulation and propulsion systems.

- Research and Development: Enabling scientific experiments and studies in fields such as physics, chemistry, and material science.

### 19.1.2 Economic Impact of Vacuum Technology

Vacuum technology plays a pivotal role in driving economic growth and competitiveness across a wide range of industries. Its economic significance stems from its ability to enhance efficiency, foster innovation, and enable the development of advanced products and processes.

**1. Driving Innovation**

- **Product Development:** Vacuum technology enables the creation of cutting-edge products by providing the necessary conditions for research, development, and manufacturing in industries such as electronics, automotive, and aerospace.

- **Process Innovation:** Vacuum processes allow for the production of materials with unique properties and characteristics, leading to the development of innovative solutions in fields like nanotechnology, biotechnology, and advanced materials.

- **2. Enhancing Efficiency**

- **Manufacturing Processes:** Vacuum technology improves manufacturing efficiency by enabling precise control over process parameters, reducing cycle times, and minimizing waste and rework.

- **Energy Efficiency:** Vacuum systems are designed to operate at optimal energy levels, resulting in reduced energy consumption and lower operational costs for industries such as semiconductor manufacturing, where energy-intensive processes are commonplace.

- **3. Driving Competitiveness**

- **Global Markets:** Industries that leverage vacuum technology gain a competitive edge in global markets by producing high-quality products with enhanced performance and reliability.

- **Cost-Effectiveness:** Vacuum technology allows for cost-effective production methods, enabling companies to offer competitive pricing while maintaining profit margins.

- **4. Facilitating Research and Development**

- **Scientific Advancements:** Vacuum technology provides researchers with the tools and capabilities needed to conduct experiments and studies in diverse scientific fields, leading to breakthrough discoveries and technological advancements.

- **Collaborative Innovation:** Collaborations between industry, academia, and research institutions drive innovation in vacuum technology, leading to the development of new applications and solutions.

- **5. Job Creation and Economic Growth**

- **Skilled Workforce:** The demand for skilled professionals in vacuum technology-related fields, such as vacuum system design, maintenance, and operation, contributes to job creation and economic growth.

- **Supporting Industries:** Vacuum technology supports a network of ancillary industries, including vacuum pump manufacturing, vacuum gauge production, and vacuum system integration, further bolstering economic activity.

- **In summary,** the economic impact of vacuum technology extends far beyond its immediate applications, driving innovation, efficiency, and competitiveness across industries and contributing to economic growth and prosperity on a global scale.

### 19.1.3 Environmental Implications of Vacuum Technology

Vacuum technology offers several environmental benefits by minimizing waste and energy consumption in industrial processes. By creating controlled environments with reduced air pressure, vacuum technology enables more efficient and sustainable production methods, thereby mitigating the environmental impact of various industries.

**1. Waste Reduction**

- **Contaminant Removal**: Vacuum systems are used to extract contaminants and impurities from materials and products during manufacturing processes, reducing the need for additional cleaning steps and minimizing waste generation.

- **Material Conservation:** Vacuum technology enables precise material deposition and coating processes, minimizing material waste and enhancing resource utilization in industries such as thin-film deposition and semiconductor manufacturing.

**2. Energy Efficiency**

- **Optimized Processes**: Vacuum systems operate at lower pressures, requiring less energy compared to traditional atmospheric processes. This energy efficiency contributes to lower greenhouse gas emissions and reduces the overall environmental footprint of industrial operations.

- **Heat Management:** Vacuum insulation techniques help conserve energy by minimizing heat transfer, improving the efficiency of thermal systems and reducing the energy required for heating and cooling processes.

**3. Pollution Prevention**

- **Emissions Reduction:** Vacuum technology minimizes the release of harmful pollutants and emissions by creating sealed environments that prevent the escape of volatile compounds and contaminants into the atmosphere.

- **Process Optimization**: Vacuum-assisted processes often produce fewer pollutants and by-products compared to conventional methods, contributing to cleaner and more sustainable industrial practices.

**4. Sustainable Manufacturing**

- **Resource Conservation:** Vacuum technology supports sustainable manufacturing practices by reducing raw material consumption, minimizing waste generation, and extending the lifespan of equipment and components through precision processing techniques.

- **Green Product Development:** The environmental benefits of vacuum technology make it an essential tool for industries seeking to develop eco-friendly products and solutions, aligning with consumer demands for sustainability.

**In conclusion,** vacuum technology plays a crucial role in promoting environmental sustainability by minimizing waste generation, reducing energy consumption, and preventing pollution in industrial processes. Embracing vacuum technology not only improves operational efficiency but also contributes to the global effort to mitigate environmental challenges and build a more sustainable future.

## 19.1.4 Importance of Understanding Vacuum Measurements, Gauges, and Pumps

- **Optimizing System Performance:** Emphasizing the critical role of accurate vacuum measurements in ensuring efficient system operation and product quality.

- **Process Control and Monitoring:** Explaining how vacuum gauges provide real-time feedback on system conditions, enabling precise control and troubleshooting.

- **Safety Considerations:** Highlighting the importance of proper vacuum pump selection and maintenance to prevent system failures and ensure operator safety.

- **Energy Efficiency:** Discussing how understanding vacuum pumps and their operating principles can lead to energy-efficient system designs and reduced operational costs.

## 19.1.5 Importance of Vacuum Technology in Various Industries and Applications

Vacuum technology serves as a cornerstone in numerous industries, enabling a wide range of applications that rely on controlled pressure environments. Its importance spans across sectors such as:

## 1. Manufacturing and Production:

- **Semiconductor Fabrication**: Vacuum chambers are essential for processes like etching, deposition, and lithography in semiconductor manufacturing.

- **Pharmaceuticals:** Vacuum drying and freeze-drying techniques are critical for pharmaceutical production, ensuring product stability and purity.

- **Food Packaging:** Vacuum sealing extends the shelf life of food products by removing air and preventing spoilage.

- **Metallurgy:** Vacuum induction melting and casting techniques are used to produce high-purity metals and alloys with precise properties.

## 2. Research and Development:

- **Physics and Materials Science:** Vacuum chambers facilitate experiments in particle physics, surface science, and material characterization.

- **Chemistry:** Vacuum distillation and filtration techniques are employed in chemical synthesis and purification processes.

- **Space Simulation:** Vacuum chambers simulate space conditions for testing spacecraft components and materials.

## 3. Energy and Environment:

- **Solar Panel Manufacturing:** Vacuum deposition processes are used to create thin-film coatings for solar cells, improving energy conversion efficiency.

- **Wastewater Treatment:** Vacuum systems aid in dewatering sludge and removing contaminants in wastewater treatment plants.

- **Oil and Gas Extraction**: Vacuum pumps assist in extracting and refining oil and gas from underground reservoirs.

## 4. Healthcare and Biotechnology:

- **Medical Imaging:** Vacuum technology is integral to the operation of equipment like MRI machines and electron microscopes.

- **Biomedical Research:** Vacuum systems support techniques such as freeze-drying and tissue culture in biomedical laboratories.

## 5. Aerospace and Aviation:

- **Vacuum Testing:** Aerospace components undergo vacuum testing to ensure they can withstand the low-pressure conditions of space.

- **Fuel Systems:** Vacuum pumps are used in aircraft fuel systems for priming and maintaining fuel flow.

**In summary,** vacuum technology plays a fundamental role in enhancing efficiency, precision, and innovation across diverse industries and applications, driving advancements in technology, science, and manufacturing processes. Its versatility and importance continue to expand as industries evolve and seek solutions for complex challenges.

## 19.1.6 Significance of Understanding Vacuum Measurements, Gauges, and Pumps

Efficient operation of vacuum systems relies heavily on a comprehensive understanding of vacuum measurements, gauges, and pumps. Here's why this knowledge is crucial:

**1. Precision Control:**

- Vacuum measurements provide critical feedback on the pressure levels within a system, allowing operators to maintain precise control over process conditions.

- Accurate vacuum gauges ensure that pressure targets are met consistently, minimizing variations that could affect product quality or system performance.

**2. Performance Optimization:**

- Understanding different types of vacuum pumps and their characteristics enables operators to select the most suitable pump for specific applications.

- Proper pump selection and sizing optimize system performance, enhancing productivity and energy efficiency while reducing operational costs.

**3. System Safety and Reliability:**

- Inadequate vacuum levels or fluctuations can compromise the integrity of processes and equipment, leading to product defects, system failures, or safety hazards.

- Knowledge of vacuum measurements and gauges allows operators to detect abnormalities early, preventing potential damage or accidents and ensuring system reliability.

**4. Process Monitoring and Troubleshooting:**

- Vacuum measurements serve as diagnostic tools for identifying leaks, blockages, or other issues within vacuum systems.

- Timely detection and resolution of problems based on gauge readings prevent downtime and production losses, maintaining operational continuity.

**5. Cost-Efficient Operation:**

- Efficient vacuum system operation minimizes energy consumption, reduces maintenance requirements, and prolongs equipment lifespan, resulting in cost savings over the long term.

- Proper understanding of vacuum technology helps optimize resource utilization and maximize return on investment in vacuum equipment and processes.

**In summary,** proficiency in vacuum measurements, gauges, and pumps is essential for achieving optimal performance, reliability, and safety in vacuum system operation. It empowers operators to maintain precise control, optimize processes, troubleshoot issues effectively, and operate vacuum systems efficiently, ultimately contributing to overall productivity and competitiveness.

## 19.2 Units of Pressure and Vacuum

## 19.2.1 Explanation of common units used to measure pressure and vacuum, including:

**Pascal (Pa)**

The Pascal (Pa) is the standard unit for measuring pressure in the International System of Units (SI). It is named after the French mathematician, physicist, and philosopher Blaise Pascal.

**Brief Overview:**

- The Pascal represents one Newton per square meter ($N/m^2$), which is the amount of force applied over a unit area.

- It is commonly used in various fields such as physics, engineering, meteorology, and fluid dynamics to quantify pressure.

- In vacuum technology, Pascals are frequently employed to measure the level of vacuum within a system. Lower Pascal values indicate higher levels of vacuum, while higher values signify greater pressure.

**Applications:**

- Vacuum systems often use Pascals to specify pressure levels, with measurements ranging from atmospheric pressure (approximately 101,325 Pa) down to ultra-high vacuum levels (e.g., in the range of 10^-9 Pa).

- Pascal measurements are crucial for ensuring the proper functioning of vacuum processes and equipment, as they provide accurate feedback on pressure conditions within a system.

- In scientific research, Pascals are used to quantify pressure changes in experiments, analyze fluid behavior, and study phenomena such as atmospheric pressure variations.

  ### Significance:

  The Pascal offers a standardized and universally recognized unit for measuring pressure, facilitating communication and comparison of pressure values across different applications and industries.

- Understanding Pascals is essential for effectively managing and controlling pressure-related parameters in various processes, ensuring safety, efficiency, and quality in engineering and scientific endeavors.

In vacuum technology, the Pascal serves as a fundamental unit for quantifying pressure levels, enabling precise measurement and control essential for the operation of vacuum systems and processes.

- ### Torr

The Torr is a unit of pressure named after the Italian physicist Evangelista Torricelli, who invented the mercury barometer in 1643. It is commonly used in vacuum technology and is equivalent to 1/760th of a standard atmospheric pressure at sea level, which is approximately 101,325 Pascals (Pa).

**Brief Overview:**

- One Torr is equal to 1/760th of atmospheric pressure at sea level, which is approximately 101,325 Pascals (Pa) or 760 millimeters of mercury (mmHg).

- It is often used to express low-pressure conditions in vacuum systems, particularly in applications where precise control and measurement of pressure are essential.

- The Torr is widely used in vacuum technology, ranging from rough vacuum pressures (e.g., a few hundred Torr) to high vacuum pressures (e.g., below 1 Torr).

**Applications:**

- Torr measurements are commonly used in vacuum systems for processes such as thin-film deposition, semiconductor manufacturing, and analytical instrumentation.

- In scientific research, Torr measurements are crucial for experiments requiring controlled pressure environments, such as in physics, chemistry, and materials science.

- Vacuum pumps and gauges are calibrated to provide pressure readings in Torr, allowing engineers and scientists to monitor and adjust pressure levels accurately.

**Significance:**

- The Torr provides a convenient and widely accepted unit for quantifying low-pressure conditions, particularly in vacuum-related applications.

- Understanding Torr measurements is essential for engineers and scientists working with vacuum systems, as it allows for precise control and optimization of pressure conditions.

- Conversion between Torr and other pressure units, such as Pascals or atmospheres, is important for ensuring consistency and compatibility across different measurement systems and standards.

**In summary**, the Torr is a valuable unit of pressure measurement in vacuum technology, offering a standardized and practical way to quantify low-pressure conditions essential for various industrial, scientific, and research applications.

- **Atmosphere (atm)**

The atmosphere (atm) is a unit of pressure commonly used to measure atmospheric pressure, particularly at sea level. It represents the average pressure exerted by Earth's atmosphere at sea level, which is approximately 101,325 Pascals (Pa) or 1,013.25 millibars (mbar).

**Brief Overview:**

- One atmosphere is defined as the pressure exerted by a column of mercury 760 millimeters (mm) in height at 0°C (32°F) and standard gravity (9.80665 m/s^2).

- Atmospheric pressure varies with altitude; it decreases as altitude increases due to the decreasing density of the air molecules.

- At sea level, the standard atmospheric pressure is approximately 1 atm, equivalent to 101,325 Pascals (Pa) or 14.7 pounds per square inch (psi).

**Applications:**

- Atmospheric pressure measurements are used in various fields, including meteorology, aviation, and scuba diving, to assess weather conditions, aircraft performance, and diving safety.

- In industrial applications, atmospheric pressure is often used as a reference point for pressure measurements in vacuum systems, where pressures are typically expressed relative to atmospheric pressure (e.g., 0.1 atm for 90% vacuum).

**Significance:**

- The atmosphere provides a convenient reference point for pressure measurements, particularly in applications related to Earth's atmosphere and its effects on human activities.

- Understanding atmospheric pressure is crucial for a wide range of activities, from weather forecasting and aviation to industrial processes and scientific research.

- Conversions between atmospheres and other pressure units, such as Pascals or Torr, are important for ensuring consistency and compatibility across different measurement systems and standards.

**In summary**, the atmosphere is a fundamental unit of pressure measurement, representing the pressure exerted by Earth's atmosphere at sea level. It serves as a standard reference point for pressure measurements in various fields and is essential for understanding atmospheric dynamics and their impacts on human activities.

- **Bar**

The bar (symbol: bar) is a unit of pressure commonly used in various applications, representing approximately the average atmospheric pressure at sea level. One bar is equivalent to 100,000 Pascals (Pa), or 100 kilopascals (kPa).

**Brief Overview:**

- The bar is a metric unit of pressure derived from the CGS (centimeter-gram-second) system, where 1 bar is equal to $10^5$ dynes per square centimeter.

- It provides a convenient measure of pressure, especially in scientific, engineering, and industrial contexts where pressure values are often relatively high.

- The bar is often used in meteorology, engineering, and manufacturing, as well as in areas such as scuba diving, where pressure measurements are essential for safety and performance.

**Applications:**

- In meteorology, atmospheric pressure is frequently measured in bars, with standard atmospheric pressure at sea level being approximately 1.01325 bars.

- In industrial settings, such as manufacturing and chemical processing, pressure gauges and instruments often display pressure values in bars for monitoring and control purposes.

- Barometric pressure readings are used in aviation for weather forecasting, flight planning, and aircraft performance calculations.

**Significance:**

- The bar provides a straightforward and widely accepted unit of pressure measurement, particularly in contexts where pressures are relatively high compared to atmospheric pressure.

- It facilitates easy conversion between different pressure units, such as Pascals, atmospheres, and millibars, which is essential for interoperability and standardization in scientific and industrial applications.

- Understanding pressure measurements in bars is crucial for ensuring the safety, efficiency, and accuracy of various processes and systems across different industries and disciplines.

In summary, the bar is a metric unit of pressure commonly used in scientific, engineering, and industrial contexts to measure and quantify pressure values. It serves as a practical and versatile unit for expressing pressure measurements and plays a vital role in a wide range of applications, from weather forecasting to industrial process control.

- **Millibar (mbar)**

The millibar (mbar) is a metric unit of pressure commonly used in meteorology and other atmospheric sciences. One millibar is equal to one one-thousandth of a bar, or

100 Pascals (Pa). It is often used to measure atmospheric pressure, especially in weather forecasting and aviation.

**Conversion between different units to facilitate understanding and communication in vacuum technology.**

To convert between different pressure units, you can use conversion factors. Here are the conversions for the units mentioned:

Pascal (Pa) to Torr:

1 Pa = 0.00750061683 Torr

Pascal (Pa) to Atmosphere (atm):

1 Pa = 9.8692 x 10^-6 atm

Pascal (Pa) to Bar:

1 Pa = 1 x 10^-5 bar

Pascal (Pa) to Millibar (mbar):

1 Pa = 0.01 mbar

You can use these conversion factors to convert between any of the mentioned pressure units. For example, to convert from Pascal to Torr:

Torr=Pa×0.00750061683

here's how to convert from each unit to Pascal:

1. Torr to Pascal: Pa=Torr×133.322Pa

2. Atmosphere (atm) to Pascal: Pa=atm×101325

3. Bar to Pascal: Pa=bar×100000Pa

4. Millibar (mbar) to Pascal: Pa=mbar×100

These formulas allow you to convert from each unit to Pascal.

**Note:**

1 Torr is equivalent to 1 mm Hg.

1 atm is equivalent to 760 mm Hg.

## 19.3 Types of Vacuum Measuring Gauges

1. **Pirani Gauge**:

   - **Operating Principle**: Measures pressure by sensing the change in thermal conductivity of the gas as pressure decreases.

   - **Operation:** Utilizes a heated wire exposed to the vacuum, where the heat transfer varies with pressure.

   - **Suitable Pressure Range:** Typically used for measuring pressure in the medium to high vacuum range ($10^{-3}$ to 1000 Torr).

2. **Thermocouple Gauge**:

   - **Operating Principle:** Measures pressure by detecting the thermal conductivity of the gas.

   - **Operation:** Utilizes a thermocouple placed within the vacuum system, where heat transfer varies with pressure.

   - **Suitable Pressure Range:** Commonly used for measuring pressure in the low to medium vacuum range ($10^{-3}$ to $10^{-9}$ Torr).

3. **Capacitance Manometer**:

   - **Operating Principle:** Measures pressure by sensing the change in capacitance due to deflection of a diaphragm.

   - **Operation:** Utilizes a diaphragm exposed to the vacuum, where deflection varies with pressure.

   - **Suitable Pressure Range:** Ideal for measuring pressure in the high vacuum range ($10^{-3}$ to $10^{-9}$ Torr).

4. **Ionization Gauge**:

   - **Operating Principle:** Measures pressure by detecting the ions produced when gas molecules are bombarded with electrons.

   - **Operation:** Utilizes an ion collector and emits electrons to ionize gas molecules.

   - **Suitable Pressure Range:** Typically used for measuring pressure in the high to ultra-high vacuum range ($10^{-3}$ to $10^{-12}$ Torr).

5. **Cold Cathode Gauge**:

- **Operating Principle:** Measures pressure by ionizing gas molecules using a high voltage electric field.

- **Operation:** Utilizes a cold cathode and anode within a vacuum chamber to generate ions.

- **Suitable Pressure Range:** Used for measuring pressure in the ultra-high vacuum range ($10^{-7}$ to $10^{-12}$ Torr).

6. **Penning Gauge**:

- **Operating Principle:** Measures pressure by detecting the ionization current produced in a magnetic field.

- **Operation:** Utilizes a cold cathode and a magnetic field to produce ions and measure the resulting current.

- **Suitable Pressure Range:** Typically used for measuring pressure in the ultra-high vacuum range ($10^{-3}$ to $10^{-9}$ Torr).

## 19.4 Vacuum Gauge Ranges and Applications

Each type of vacuum gauge has a specific pressure range in which it operates effectively, making it crucial to choose the right gauge for the desired pressure measurement. Here's a summary of the pressure ranges each type of vacuum gauge can measure effectively:

- **Pirani Gauge:**

  - Effective Range: Medium to high vacuum range ($10^{-3}$ to 1000 Torr).

  - Thermocouple Gauge:

  - Effective Range: Low to medium vacuum range ($10^{-3}$ to $10^{-9}$ Torr).

  - **Capacitance Manometer:**

  Effective Range: High vacuum range ($10^{-3}$ to $10^{-9}$ Torr).

- **Ionization Gauge:**

  - Effective Range: High to ultra-high vacuum range ($10^{-3}$ to $10^{-12}$ Torr).

- **Cold Cathode Gauge:**

  Effective Range: Ultra-high vacuum range ($10^{-7}$ to $10^{-12}$ Torr).

- **Penning Gauge:**

    Effective Range: Ultra-high vacuum range (10^-3 to 10^-9 Torr).


Understanding these pressure ranges helps in selecting the appropriate gauge for accurate pressure measurement in vacuum systems across various operational conditions.

**Examples of Applications for Each Vacuum Gauge Type**

1. **Pirani Gauge**:

    **Applications**: Used in research labs, industrial vacuum processes like coating and drying, and semiconductor manufacturing for monitoring vacuum levels during deposition processes.

2. **Thermocouple Gauge**:

    **Applications**: Commonly employed in research labs, vacuum furnaces, and pharmaceutical industry for monitoring vacuum levels in freeze-drying processes and vacuum packaging.


3. **Capacitance Manometer**:

    **Applications**: Widely used in semiconductor manufacturing for monitoring vacuum levels in lithography and etching processes, as well as in research labs and vacuum systems where high accuracy is required.

4. **Ionization Gauge**:

    **Applications**: Essential in ultra-high vacuum systems used in accelerator facilities, space simulation chambers, and surface science experiments for precise pressure measurements.

5. **Cold Cathode Gauge**:

    **Applications**: Utilized in semiconductor manufacturing for monitoring vacuum levels in thin-film deposition processes such as physical vapor deposition (PVD) and chemical vapor deposition (CVD).

6. **Penning Gauge**:

   **Applications**: Commonly found in research labs, vacuum systems for particle accelerators, and space simulation chambers for accurate pressure measurements in ultra-high vacuum environments.

Understanding these applications helps in selecting the appropriate vacuum gauge for specific industrial processes and research activities where precise pressure measurement is critical.

## 19.5 Vacuum Pump Types and Their Uses

Vacuum pumps are essential components in cleanroom environments, playing a critical role in creating and maintaining the required vacuum conditions. Understanding the different types of vacuum pumps and their operating principles is fundamental for efficient cleanroom operations. In this subchapter, we will explore various types of vacuum pumps commonly used in cleanroom settings and their applications based on factors such as pumping speed, ultimate vacuum, and compatibility with different gases.

1. **Rotary Vane Pumps:**

   - Operating Principle: Rotary vane pumps utilize rotating vanes to create a vacuum by compressing gas in a chamber and then expelling it through an exhaust port. These pumps are positive displacement pumps, meaning they trap gas molecules and transfer them from the inlet to the outlet.

   - Applications: Rotary vane pumps are versatile and widely used in cleanrooms for general-purpose vacuum applications, including vacuum drying, vacuum filtration, and vacuum distillation. They are suitable for achieving moderate vacuum levels and are compatible with a wide range of gases.

2. **Diaphragm Pumps:**

   - Operating Principle: Diaphragm pumps use a flexible diaphragm to generate vacuum by alternately compressing and expanding a chamber. This movement creates suction and expulsion cycles, effectively evacuating gas from the system.

   - Applications: Diaphragm pumps are ideal for applications requiring oil-free vacuum, such as laboratory and cleanroom operations where contamination is a concern. They are commonly used for vacuum filtration, solvent evaporation, and sample concentration.

3. **Scroll Pumps:**

- Operating Principle: Scroll pumps employ two interleaved spiral scrolls to compress gas and create a vacuum. As the scrolls orbit, they trap and progressively compress gas, leading to the generation of vacuum.

- Applications: Scroll pumps are known for their quiet operation, reliability, and oil-free design, making them suitable for cleanroom applications where contamination control is crucial. They are often used in semiconductor manufacturing, analytical instrumentation, and research laboratories.

4. **Turbo Pumps:**

- Operating Principle: Turbo pumps utilize high-speed rotating blades to create a vacuum by transferring gas molecules from the inlet to the exhaust at high velocities. These pumps achieve high vacuum levels by generating continuous gas flow and effectively removing gas molecules from the system.

- Applications: Turbo pumps are essential for achieving ultra-high vacuum levels required in advanced cleanroom processes such as thin-film deposition, ion implantation, and surface analysis techniques like X-ray photoelectron spectroscopy (XPS) and secondary ion mass spectrometry (SIMS).

5. **Cryogenic Pumps:**

- Operating Principle: Cryogenic pumps operate by condensing gas molecules at extremely low temperatures, typically using liquid nitrogen or helium. As gas molecules come into contact with the cryogenic surfaces, they freeze and are removed from the system.

- Applications: Cryogenic pumps are utilized in cleanrooms for pumping inert gases and achieving high vacuum levels in applications such as semiconductor manufacturing, particle accelerators, and vacuum chambers for space simulation.

Explanation of Applications:

Each type of vacuum pump offers distinct advantages and is suited for specific cleanroom applications based on factors such as pumping speed, ultimate vacuum level, gas compatibility, and contamination control requirements. Understanding the operating principles and applications of different vacuum pump types is essential for selecting the most suitable pump for a particular cleanroom process or application.

This subchapter provides an overview of the most common vacuum pump types used in cleanrooms and their respective applications, helping readers make informed decisions regarding vacuum pump selection and integration into cleanroom systems.

## 19.6 Considerations for Selecting Vacuum Equipment

When it comes to selecting vacuum equipment for cleanroom applications, several factors must be carefully considered to ensure optimal performance and compatibility with specific process requirements. In this section, we will explore key considerations for selecting vacuum gauges and pumps tailored to the needs of cleanroom environments.

1. **Required Pressure Range:**

   The first consideration in selecting vacuum equipment is determining the required pressure range for the application. Different processes may require varying levels of vacuum, ranging from high vacuum (ultra-high vacuum) to rough vacuum. Understanding the desired pressure range is essential for choosing the appropriate vacuum gauge and pump capable of achieving and maintaining the desired vacuum level.

2. **Pumping Speed:**

   Pumping speed refers to the rate at which a vacuum pump can evacuate gas from a system. It is crucial to assess the pumping speed requirements based on the volume of gas to be evacuated and the desired evacuation time. Selecting a vacuum pump with an adequate pumping speed ensures efficient gas removal and timely attainment of the desired vacuum level.

3. **Compatibility with Process Gases:**

   Another critical factor is the compatibility of vacuum equipment with process gases. Certain gases may react with pump components or lubricants, leading to contamination or degradation of pump performance. It is essential to consider the chemical compatibility of vacuum pumps and gauges with process gases to prevent contamination and ensure reliable operation.

4. **Cost and Maintenance Requirements:**

Cost and maintenance considerations play a significant role in selecting vacuum equipment for cleanroom applications. While initial equipment cost is a factor, it is equally important to evaluate long-term maintenance requirements, including service intervals, replacement parts availability, and associated costs. Choosing reliable vacuum equipment with low maintenance requirements can contribute to overall operational efficiency and cost-effectiveness.

5. **Integration with Existing Vacuum Systems:**

When selecting vacuum equipment, compatibility and integration with existing vacuum systems must be taken into account. It is essential to ensure that new vacuum gauges and pumps can seamlessly integrate with the existing infrastructure, including vacuum lines, valves, and control systems. Compatibility ensures smooth operation and facilitates system upgrades or expansions in the future.

By carefully considering these factors when selecting vacuum equipment, cleanroom operators can ensure the optimal performance, reliability, and compatibility of vacuum systems with specific process requirements. Making informed decisions based on pressure range, pumping speed, gas compatibility, cost, maintenance, and integration considerations is essential for achieving efficient and reliable vacuum operations in cleanroom environments.

## 19.7 Case Studies and Practical Examples

In this section, we delve into real-world examples that highlight the application of different vacuum measurement techniques and pump technologies across various industries. Through case studies, we illustrate the significance of selecting the right vacuum equipment for achieving optimal performance and efficiency in specific applications.

1. **Semiconductor Industry:**

In the semiconductor industry, precise vacuum control is critical for fabricating integrated circuits with nanoscale features. Case studies showcase the use of capacitance manometers and ionization gauges for accurate pressure measurement during wafer processing. Additionally, examples demonstrate the deployment of dry vacuum pumps in semiconductor cleanrooms to maintain ultra-clean environments and prevent contamination during chip manufacturing.

2. **Pharmaceutical Manufacturing:**

   Pharmaceutical manufacturing facilities require vacuum systems for processes such as drying, distillation, and solvent recovery. Practical examples highlight the use of rotary vane pumps and vacuum ovens in pharmaceutical laboratories to facilitate efficient solvent removal and product drying. Case studies emphasize the importance of vacuum equipment in ensuring product quality, compliance with regulatory standards, and operational efficiency in pharmaceutical production.

3. **Research Laboratories:**

   Research laboratories often utilize a wide range of vacuum equipment for experiments and scientific investigations. Case studies illustrate the use of turbo molecular pumps and cryogenic pumps in high-vacuum chambers for studying material properties and conducting particle physics experiments. Practical examples demonstrate how the selection of appropriate vacuum technology enhances experimental reproducibility, sensitivity, and data accuracy in research settings.

4. **Food Packaging Industry:**

   Vacuum packaging is a common technique used in the food industry to extend the shelf life of perishable products and maintain product freshness. Case studies examine the deployment of vacuum sealers and vacuum chambers in food packaging facilities for removing air from packaging containers and preserving food quality. Practical examples showcase the role of vacuum technology in reducing food waste, improving product shelf life, and enhancing food safety standards.

5. **Analytical Instrumentation:**

   - Analytical instruments such as mass spectrometers and electron microscopes rely on vacuum systems to create low-pressure environments for accurate sample analysis. Case studies highlight the integration of high-performance vacuum gauges and pumps in analytical instruments to achieve precise pressure control and measurement. Practical examples demonstrate the impact of vacuum equipment on analytical instrument sensitivity, resolution, and detection limits, enabling advanced scientific research and analysis.

Through these case studies and practical examples, we gain insights into the diverse applications of vacuum technology across industries and the importance of selecting the right vacuum equipment for specific operational requirements. By understanding real-world scenarios and their outcomes, stakeholders can make informed decisions regarding vacuum system design, implementation, and optimization to enhance productivity, quality, and innovation in their respective fields.

# Chapter 20: Technical Semiconductor Cleanroom Tools/Devices Experience

## 20.1 Optical vs. tactile profiler?

Optical and tactile profilers are both instruments used for surface metrology to measure the topography and surface characteristics of materials. However, they differ in their measurement techniques and the types of data they provide.

1. **Optical Profiler:**

   - **Measurement Technique:** Optical profilers use light to measure surface topography. Common techniques include interferometry and focus variation.

   - **Contact:** Non-contact measurement. The instrument uses light waves to interact with the surface, making it suitable for delicate or sensitive samples.

   - **Speed:** Generally faster than tactile profilers for obtaining surface measurements.

   - **Applications:** Well-suited for measuring surface features with sub-nanometer to micrometer scale roughness, as well as determining features such as step heights, surface waviness, and form errors.

   - **Usage of optical profiler:**

     - Step height

     - Roughness

     - Critical Dimensions analysis

     - Film Thickness

     - Film Conformality

2. **Tactile Profiler (Stylus Profiler):**

   - **Measurement Technique:** Tactile profilers use a physical stylus or probe that makes direct contact with the surface to measure height variations. The stylus traces the surface contours, and the vertical movement is recorded.

   - **Contact:** Requires physical contact with the surface, making it less suitable for delicate or easily damaged samples.

- **Speed:** Generally slower than optical profilers because the stylus has to physically traverse the surface.

- **Applications:** Effective for measuring surface roughness, texture, and step heights. Tactile profilers are often used when precise vertical resolution and accurate height measurements are crucial.

**In summary**, the main difference lies in the measurement technique and whether contact with the surface is made. Optical profilers are non-contact instruments that use light to measure, whereas tactile profilers use a physical stylus in direct contact with the surface. The choice between them depends on the specific requirements of the application, the nature of the sample, and the desired level of precision.

**Comparison between optical profiler and tactile/stylus profiler?**

1. **Measurement Technique:**

   - **Optical Profiler:** Uses light-based techniques such as interferometry or focus variation for non-contact surface measurements.

   - **Tactile/Stylus Profiler:** Employs a physical stylus or probe that makes direct contact with the surface, tracing its contours.

2. **Contact:**

   - **Optical Profiler:** Non-contact, making it suitable for delicate or sensitive samples.

   - **Tactile/Stylus Profiler:** Requires physical contact with the surface, which may not be suitable for easily damaged or sensitive samples.

3. **Speed:**

   - **Optical Profiler:** Generally faster due to non-contact measurements.

   - **Tactile/Stylus Profiler:** Slower as the stylus physically traverses the surface.

4. **Applications:**

   - **Optical Profiler:** Well-suited for measuring features with sub-nanometer to micrometer scale roughness, step heights, surface waviness, and form errors.

   - **Tactile/Stylus Profiler:** Effective for measuring surface roughness, texture, step heights, and precise vertical resolution.

5. **Sample Characteristics:**

   - **Optical Profiler:** Suitable for a wide range of samples, including delicate or transparent materials.

   - **Tactile/Stylus Profiler:** May be limited in use for softer materials or those prone to damage during contact.

6. **Ease of Use:**

   - **Optical Profiler:** Generally user-friendly, with non-contact measurements being easier to set up.

   - **Tactile/Stylus Profiler:** Requires careful setup to avoid damaging the sample and the stylus needs to be chosen based on the surface characteristics.

7. **Vertical Resolution:**

   - **Optical Profiler:** Can achieve high vertical resolution.

   - **Tactile/Stylus Profiler:** Offers precise vertical resolution, often essential for some applications.

8. **Cost:**

   - **Optical Profiler:** Typically more expensive due to advanced technology.

   - **Tactile/Stylus Profiler:** Generally more affordable.

Choosing between these profilers depends on the specific requirements of your application, the nature of the sample, and the level of precision needed for surface measurements.

**So, why do we  need both in one cleanroom?**

Having both optical and tactile/stylus profilers in a cleanroom can offer a comprehensive approach to surface metrology, providing a more versatile capability to meet the varied needs of semiconductor or precision manufacturing environments. Here are a few reasons why you might consider having both types of profilers:

1. **Diverse Surface Characteristics:**

   - **Optical Profiler:** Ideal for measuring features with sub-nanometer to micrometer scale roughness and complex surface structures.

   - **Tactile/Stylus Profiler:** Effective for precise measurements of surface roughness, texture, and step heights.

2. **Measurement Flexibility:**

   - Use the **Optical Profiler** for non-contact measurements on delicate or sensitive samples.

   - Employ the **Tactile/Stylus** *Profiler* for direct contact measurements when high precision and detailed vertical resolution are crucial.

3. **Applications with Unique Requirements:**

   Certain applications may demand the advantages of both profilers for a more thorough characterization of surfaces, especially in semiconductor manufacturing where a wide range of materials and structures are encountered.

4. **Process Optimization:**

   Combining the strengths of both profilers allows for a more comprehensive understanding of the surfaces involved in the manufacturing process, facilitating optimization and quality control.

5. **Material Compatibility:**

   Some materials or surfaces may be better suited to one type of profiler over the other. Having both options ensures compatibility with a broader range of materials.

6. **Cost-Effective Approach:**

   While optical profilers can be more expensive, a combination of both profilers allows for cost-effective solutions that address specific measurement needs without compromising accuracy or efficiency.

**In summary**, having both optical and tactile/stylus profilers in a cleanroom provides versatility, allowing you to choose the most appropriate tool for specific applications. This comprehensive approach enhances the capability to meet the diverse surface metrology requirements encountered in advanced manufacturing environments.

## 20.2 Ellipsometer

Ellipsometry is a powerful technique used in cleanroom environments for characterizing thin films and surfaces with high precision. This subchapter explores the principles of ellipsometry and its diverse applications in semiconductor fabrication.

**Principles of Ellipsometry:**

Ellipsometry measures changes in the polarization state of light reflected from a sample surface to extract information about its optical properties. By analyzing the changes in the amplitude and phase of the reflected light, ellipsometers can determine key parameters such as film thickness, refractive index, and optical constants.

**Key Components of an Ellipsometer:**

1. **Light Source:**

   Ellipsometers typically use a monochromatic or broadband light source, such as a laser or white light, to illuminate the sample surface.

2. **Polarizer and Analyzer:**

   These optical components control the polarization state of the incident and reflected light, allowing for precise measurement of the ellipsometric parameters.

3. **Detector:**

   Detects and records the intensity of light reflected from the sample surface at different polarization states.

4. **Sample Stage:**

   Supports the sample and allows for precise positioning and alignment during measurements.

**Applications of Ellipsometry in Cleanroom Environments:**

1. **Thin Film Characterization:**

   Ellipsometry is widely used to measure the thickness and optical properties of thin films deposited on semiconductor substrates. This includes dielectric layers, metal films, and organic coatings used in various fabrication processes.

2. **Material Analysis:**

   Ellipsometry provides valuable insights into the composition and structure of materials in cleanroom environments. It can identify surface contaminants, interface roughness, and layer composition with high sensitivity.

3. **Process Monitoring and Control:**

Ellipsometry enables real-time monitoring of thin film deposition processes, allowing for precise control of film thickness and uniformity. This ensures the reproducibility and quality of semiconductor devices manufactured in cleanroom facilities.

4. **Surface Passivation:**

Ellipsometry is employed to evaluate the effectiveness of surface passivation techniques used to reduce surface recombination in semiconductor devices. It assesses the quality and uniformity of passivation layers critical for enhancing device performance and reliability.

5. **Metrology and Quality Assurance:**

Ellipsometry serves as a non-destructive metrology tool for quality assurance in cleanroom operations. It verifies adherence to specifications, validates process parameters, and ensures the consistency of thin film properties across production batches.

6. **Research and Development:**

Ellipsometry supports R&D activities in cleanroom environments by providing detailed optical characterization of novel materials and device structures. It aids in the development of advanced semiconductor technologies and next-generation electronic devices.

In summary, ellipsometry plays a vital role in semiconductor cleanrooms by enabling precise and non-destructive characterization of thin films and surfaces. Its applications span across thin film deposition, material analysis, process control, and quality assurance, contributing to the advancement of semiconductor technology and device fabrication processes.

## 20.3 Stress gauge

In semiconductor fabrication, the precise measurement and control of stress within thin films are crucial for ensuring the reliability and performance of integrated circuits. This subchapter delves into the principles of stress gauges and their significance in cleanroom environments.

**Principles of Stress Gauges:**

A stress gauge, also known as a stress sensor or stress measurement system, is a specialized tool used to quantify the mechanical stress present in thin films deposited on semiconductor substrates. These gauges employ various techniques to measure stress, including wafer curvature, X-ray diffraction, and optical methods such as spectroscopic reflectometry.

**Key Components of a Stress Gauge:**

1. **Probe or Sensor:**

   The component directly responsible for detecting stress-induced changes in the sample. This may include a cantilever beam, piezoresistive sensor, or optical interferometer.

2. **Measurement System:**

   This includes the electronics and software necessary to interpret the signals from the probe and calculate the stress levels in the thin film.

3. **Sample Stage:**

   Supports the semiconductor wafer or substrate during stress measurements and allows for precise positioning and alignment.

**Applications of Stress Gauges in Cleanroom Environments:**

1. **Thin Film Characterization:**

   Stress gauges provide critical insights into the mechanical properties of thin films deposited during semiconductor fabrication processes. They measure stress levels resulting from deposition techniques such as physical vapor deposition (PVD), chemical vapor deposition (CVD), and plasma-enhanced chemical vapor deposition (PECVD).

2. **Process Optimization:**

   By monitoring stress levels in real-time, stress gauges facilitate the optimization of thin film deposition processes to minimize stress-induced defects and ensure uniform film properties across wafers. This aids in enhancing device yield and reliability.

3. **Material Selection:**

   Stress gauges assist cleanroom engineers and researchers in evaluating the mechanical compatibility of materials used in semiconductor fabrication. They determine the suitability of thin film materials for specific applications based on their stress characteristics.

4. **Failure Analysis:**

In the event of device failure or reliability issues, stress gauges play a crucial role in identifying the root causes by pinpointing stress-induced defects or structural weaknesses in thin film layers. This information informs failure analysis investigations and corrective actions.

5. **Quality Assurance:**

Stress measurements performed using gauges contribute to quality assurance efforts in cleanroom operations by verifying adherence to stress specifications and ensuring the consistency of thin film properties across production runs. This aids in maintaining product quality and reliability.

6. **Research and Development:**

Stress gauges support R&D activities in cleanroom environments by facilitating the characterization of novel materials and processes. They enable researchers to explore innovative thin film deposition techniques and develop advanced semiconductor technologies with optimized stress profiles.

In summary, stress gauges are indispensable tools in semiconductor cleanrooms, providing valuable insights into the mechanical properties of thin films and supporting process optimization, material selection, failure analysis, quality assurance, and research and development efforts. Their precise measurements contribute to the advancement of semiconductor technology and the fabrication of reliable integrated circuits.

## 20.4 Wire Bonder

Wire bonding is a fundamental process in semiconductor packaging and assembly, essential for establishing electrical connections between integrated circuits (ICs) and the package substrate. This subchapter explores the role of wire bonders in cleanroom environments and their significance in semiconductor manufacturing.

**Principles of Wire Bonding:**

Wire bonding is a semiconductor packaging technique that involves the attachment of thin metal wires (typically gold or aluminum) between the bonding pads on the surface of an IC chip and the leads of the package substrate. The bonding process is typically performed using specialized equipment known as wire bonders, which apply heat, pressure, and ultrasonic energy to create metallurgical bonds between the wire and the bonding pads.

**Key Components of a Wire Bonder:**

1. **Bonding Head:**

   The component responsible for positioning and securing the wire onto the bonding pad using heat, pressure, and ultrasonic energy.

2. **Wire Feed Mechanism:**

   Supplies the bonding wire from a spool to the bonding head, ensuring precise positioning and tension control.

3. **Bonding Tool:**

   A fine-tipped tool or capillary that grasps and guides the wire during the bonding process, facilitating accurate wire placement and alignment.

4. **Workstage:**

   Supports the semiconductor device or package substrate during wire bonding, providing stability and precise positioning.

**Applications of Wire Bonders in Cleanroom Environments:**

1. **Integrated Circuit Packaging:**

   Wire bonders play a crucial role in the assembly and packaging of ICs, enabling the creation of electrical interconnections between semiconductor chips and the package substrate. This process is essential for establishing reliable electrical pathways and ensuring the functionality of semiconductor devices.

2. **Die Attach:**

   In semiconductor packaging, wire bonders are often used in conjunction with die attach equipment to secure the semiconductor chip onto the package substrate before wire bonding. This ensures proper alignment and positioning of the chip, facilitating accurate wire bonding and optimizing device performance.

3. **Package Sealing:**

   After wire bonding is complete, semiconductor packages may undergo additional sealing or encapsulation processes to protect the bonded wires and semiconductor chip from environmental factors such as moisture, dust, and mechanical stress. Wire bonders aid in the sealing process by providing electrical continuity and mechanical stability to the wire bonds.

4. **Quality Assurance:**

Wire bonders contribute to quality assurance efforts in cleanroom environments by ensuring the integrity and reliability of wire bonds through precise control of bonding parameters such as bond force, ultrasonic energy, and bond time. This helps to minimize defects such as bond voids, wire lifts, and bond ball cracks, enhancing the overall quality of semiconductor devices.

5. **Failure Analysis:**

In the event of device failure or reliability issues, wire bonders support failure analysis investigations by enabling the examination of wire bonds for signs of defects or anomalies. This information helps engineers identify root causes of failure and implement corrective actions to improve device reliability.

6. **Research and Development:**

Wire bonders are utilized in R&D activities within cleanroom environments to explore novel bonding techniques, materials, and processes aimed at improving wire bonding efficiency, reliability, and performance. Research efforts may focus on advancing wire bonding technologies for emerging semiconductor applications such as microelectronics, optoelectronics, and power devices.

In summary, wire bonders are indispensable tools in semiconductor cleanrooms, playing a critical role in IC packaging and assembly processes. Their precise control and reliability contribute to the production of high-quality semiconductor devices and facilitate ongoing innovation in semiconductor packaging technology.

## 20.5 Wafer Bonder

Wafer bonding is a pivotal process in semiconductor manufacturing, enabling the integration of multiple wafers or chips to create complex semiconductor structures. This subchapter delves into the significance of wafer bonders in cleanroom environments and their diverse applications in semiconductor fabrication.

**Principles of Wafer Bonding:**

Wafer bonding involves the fusion of two or more semiconductor substrates, typically silicon wafers, to form a permanent bond at the atomic or molecular level. This bonding process is critical for the fabrication of advanced semiconductor devices, such as MEMS (Micro-Electro-Mechanical Systems), CMOS (Complementary Metal-Oxide-Semiconductor) image sensors, and 3D integrated circuits.

**Key Components of a Wafer Bonder:**

1. **Bonding Chamber:**

   The enclosure where the wafer bonding process takes place, providing a controlled environment for optimal bonding conditions.

2. **Alignment Stage:**

   Allows precise alignment and positioning of the wafers before bonding, ensuring accurate alignment of features and patterns.

3. **Heating Element:**

   Applies heat to the wafers to facilitate bonding, promoting surface activation and atomic diffusion between the bonded substrates.

4. **Pressure Control System:**

   Regulates the pressure exerted on the wafers during bonding, ensuring intimate contact between the bonding surfaces and promoting bond formation.

5. **Vacuum System:**

   Creates a vacuum environment within the bonding chamber to remove air and contaminants, enhancing the quality and reliability of the wafer bond.

**Applications of Wafer Bonders in Cleanroom Environments:**

1. **MEMS Fabrication:**

   Wafer bonders play a vital role in the production of MEMS devices, facilitating the bonding of silicon wafers to create complex microstructures and functional layers. Wafer bonding techniques such as anodic bonding, fusion bonding, and adhesive bonding are used to integrate MEMS components such as sensors, actuators, and microfluidic channels.

2. **CMOS Image Sensor Integration:**

   In CMOS image sensor manufacturing, wafer bonders are employed to assemble stacked sensor wafers, enabling the integration of imaging pixels, readout circuits, and signal processing layers. Wafer bonding techniques such as direct bonding and oxide bonding are utilized to achieve high-density pixel integration and improve sensor performance.

3. **3D Integrated Circuits:**

   Wafer bonders are essential for the fabrication of 3D integrated circuits, which involve stacking and bonding multiple semiconductor wafers to create vertically interconnected device layers. Through-silicon via (TSV) technology, hybrid bonding, and wafer-to-wafer bonding are employed to establish electrical connections between stacked wafers, enabling the integration of logic, memory, and sensor components in compact form factors.

4. **Hermetic Packaging:**

   Wafer bonders are used to seal semiconductor devices in hermetic packages, protecting them from moisture, contaminants, and mechanical stress. Bonding techniques such as fusion bonding and glass frit bonding are employed to create robust seals between the semiconductor die and the package substrate, ensuring long-term device reliability in harsh environments.

5. **Wafer-Level Packaging:**

   Wafer bonders enable wafer-level packaging (WLP) processes, where semiconductor devices are encapsulated and interconnected at the wafer level before dicing. WLP techniques such as wafer-to-wafer bonding and wafer-level chip-scale packaging (WLCSP) enable the fabrication of compact, cost-effective semiconductor packages with high interconnect density and enhanced electrical performance.

In conclusion, wafer bonders are indispensable tools in semiconductor cleanrooms, supporting a wide range of applications in MEMS fabrication, CMOS image sensor integration, 3D integrated circuits, hermetic packaging, and wafer-level packaging. Their precise control, reliability, and versatility contribute to the production of advanced semiconductor devices with improved performance, functionality, and reliability.

## 20.6 Spin Coater vs. Spray Coater:
**A Comparison and Their Applications in Cleanroom Environments.**

Spin coating and spray coating are two common techniques used in cleanroom environments for applying thin films or coatings onto substrates such as silicon wafers. This subchapter explores the differences between spin coaters and spray coaters, their operating principles, and their diverse applications in semiconductor fabrication and microfabrication processes.

**Spin Coater:**

Spin coating, also known as spin deposition, is a widely used technique for applying uniform thin films or coatings onto flat substrates. The process involves dispensing a liquid precursor or solution onto the center of a spinning substrate, typically a silicon wafer, which spreads due to centrifugal force, forming a thin film with controlled thickness and uniformity.

**Operating Principle:**

1. **Dispensing:**

   The liquid precursor is dispensed onto the center of the spinning substrate using a syringe, pipette, or dispensing nozzle.

2. **Spinning:**

   The substrate is rapidly spun at high speeds, typically ranging from hundreds to thousands of revolutions per minute (RPM), causing the liquid to spread outward and form a thin film across the surface.

3. **Evaporation:**

   As the substrate continues to spin, the solvent or carrier liquid in the precursor evaporates, leaving behind a uniform thin film or coating on the substrate.

4. **Ramp Down:**

   The spinning speed is gradually decreased to stabilize the thin film, allowing it to undergo further processing or curing.

**Applications of Spin Coaters in Cleanroom Environments:**

1. **Photoresist Coating:**

   Spin coaters are commonly used in photolithography processes to apply photoresist materials onto silicon wafers, enabling the patterning of semiconductor devices and integrated circuits.

2. **Planarization:**

   Spin coating is utilized for planarizing uneven surfaces by depositing a spin-on dielectric (SOD) or spin-on glass (SOG) material, improving the topography and surface flatness of semiconductor substrates.

3. **Thin Film Deposition:**

Spin coating is employed for depositing thin films of organic semiconductors, polymers, dielectrics, and functional materials for various electronic and optoelectronic applications.

4. **Surface Modification:**

Spin coating is utilized for functionalizing surfaces with self-assembled monolayers (SAMs), surface modifiers, and thin coatings to control surface properties such as wettability, adhesion, and biocompatibility.

## Spray Coater:

Spray coating, also known as atomization coating or aerosol deposition, is a coating technique that involves atomizing a liquid precursor into fine droplets and spraying them onto a substrate surface using a nozzle or sprayer. The process enables the deposition of thin films or coatings with controlled thickness and coverage over large areas.

## Operating Principle:

1. **Atomization:**

The liquid precursor is atomized into fine droplets using a pneumatic or ultrasonic atomizer, generating a spray mist or aerosol.

2. **Spraying:**

The atomized droplets are propelled toward the substrate surface using pressurized gas or air, forming a uniform coating as they impinge and spread on the substrate.

3. **Evaporation or Curing:**

The solvent or carrier liquid in the precursor evaporates or undergoes curing, resulting in the formation of a thin film or coating on the substrate surface.

## Applications of Spray Coaters in Cleanroom Environments:

1. **Anti-reflective Coatings:**

Spray coating is utilized for depositing anti-reflective coatings (ARCs) onto optical components and photovoltaic devices to reduce surface reflections and enhance light transmission or absorption.

2. **Encapsulation Layers:**

Spray coating is employed for applying encapsulation layers or protective coatings onto electronic devices, sensors, and MEMS devices to prevent moisture ingress, corrosion, and mechanical damage.

3. **Functional Coatings:**

Spray coating is utilized for depositing functional coatings such as hydrophobic, hydrophilic, or oleophobic coatings onto surfaces to modify surface properties and enhance performance in specific applications.

4. **Barrier Films:**

Spray coating is employed for depositing barrier films or moisture barriers onto flexible substrates, films, or packaging materials to protect sensitive electronic components from environmental degradation and moisture-induced failures.

**Conclusion:**

In summary, spin coating and spray coating are versatile techniques employed in cleanroom environments for depositing thin films or coatings onto substrates in semiconductor fabrication, microfabrication, and various industries. While spin coating offers excellent control over film thickness and uniformity for small-area substrates, spray coating excels in coating large-area substrates with high throughput and efficiency. Understanding the differences and applications of spin coaters and spray coaters is essential for optimizing coating processes and achieving desired film properties in cleanroom manufacturing environments.

## 20.7 Spray Developer vs. regular developing:
### A Comparison and Their Applications in Cleanroom Environments.

In cleanroom environments, the development process is crucial for patterning photoresist materials and defining intricate features on semiconductor substrates. This subchapter explores the differences between spray developers and regular developers, their operating principles, and their diverse applications in semiconductor fabrication and microfabrication processes.

**Regular Developing:**

Regular developing, also known as immersion developing, is a conventional method for removing unexposed or soluble regions of photoresist material from a substrate surface. The process involves immersing the substrate, typically a silicon wafer, in a liquid developer solution, which selectively dissolves the unexposed or uncured photoresist, leaving behind the desired patterned features.

**Operating Principle:**

1. **Immersion:**

   The substrate with the patterned photoresist is immersed in a developer solution contained in a development bath or tank.

2. **Selective Dissolution:**

   The developer solution selectively dissolves the unexposed or soluble regions of the photoresist material, while the exposed or cross-linked regions remain intact.

3. **Agitation:**

   Agitation techniques such as stirring or ultrasonic agitation enhance the removal of dissolved photoresist material and ensure uniform development across the substrate surface.

4. **Rinsing:**

   After the desired development time, the substrate is rinsed with deionized water or a rinse solution to remove residual developer and prevent contamination.

**Applications of Regular Developing in Cleanroom Environments:**

1. **Photolithography:**

   Regular developing is commonly used in photolithography processes to define patterns on silicon wafers for the fabrication of semiconductor devices and integrated circuits.

2. **MEMS Fabrication:**

   Regular developing is employed in microelectromechanical systems (MEMS) fabrication processes to define microstructures and release sacrificial layers, enabling the fabrication of MEMS devices such as accelerometers, gyroscopes, and pressure sensors.

3. **Thin Film Patterning:**

   Regular developing is utilized for patterning thin films of photoresist, polymers, or organic materials deposited on substrates for various electronic, optical, and biomedical applications.

**Spray Developer:**

Spray development, also known as spray developing or spray rinsing, is an alternative method for developing photoresist patterns on substrates using a spray of developer solution. Unlike immersion developing, which involves submerging the substrate in a liquid bath, spray development applies the developer solution directly onto the substrate surface using a nozzle or sprayer, resulting in faster development times and reduced chemical consumption.

**Operating Principle:**

1. **Spraying:**

   The developer solution is sprayed onto the surface of the substrate using a nozzle or sprayer, forming a fine mist or spray pattern that covers the entire substrate surface.

2. **Selective Dissolution:**

   The sprayed developer solution selectively dissolves the unexposed or soluble regions of the photoresist material, similar to immersion developing, while the exposed or cross-linked regions remain unaffected.

3. **Rinsing:**

   After the development process, the substrate may be rinsed with deionized water or a rinse solution to remove residual developer and prevent contamination. In some cases, the spray developer itself may serve as the rinse solution, combining development and rinsing steps into a single process.

**Applications of Spray Developer in Cleanroom Environments:**

1. **High-Throughput Processing:**

   Spray development offers faster development times compared to immersion developing, making it suitable for high-throughput manufacturing environments where rapid processing is essential.

2. **Large-Area Substrates:**

   Spray development is particularly useful for processing large-area substrates or panels, where immersion tanks may be impractical or require significant volumes of developer solution.

3. **Reduced Chemical Consumption:**

   Spray development consumes less developer solution compared to immersion developing, resulting in reduced chemical usage and waste generation, which aligns with sustainability and environmental conservation goals.

4. **Minimized Contamination:**

   Spray development minimizes the risk of cross-contamination between substrates by applying developer solution directly onto the substrate surface without immersion in a shared bath, ensuring process cleanliness and integrity.

**Conclusion:**

In summary, both spray development and regular development are essential techniques used in cleanroom environments for patterning photoresist materials and defining microscale features on semiconductor substrates. While regular developing offers simplicity and versatility for small-scale processing, spray development provides advantages in terms of speed, efficiency, and chemical usage for high-throughput manufacturing and large-area substrate processing applications. Understanding the differences and applications of spray developers and regular developers is crucial for optimizing development processes and achieving desired patterned features in semiconductor fabrication and microfabrication processes.

# 20.8 Plasma Activator:
## Enhancing Surface Properties in Cleanroom Environments

Plasma activation, also known as plasma treatment or plasma cleaning, is a surface modification technique widely used in cleanroom environments to enhance the surface properties of materials through the interaction of energetic plasma species. This subchapter explores the principles of plasma activation, its applications in semiconductor fabrication and microfabrication processes, and its significance in achieving precise and reliable surface modifications in cleanroom environments.

**Operating Principle:**

Plasma activation involves subjecting a material surface to a low-pressure, glow discharge plasma generated within a vacuum chamber or reactor. The plasma consists of a highly energetic mixture of ions, electrons, radicals, and photons, which interact with the material surface, leading to various surface modifications and functionalization.

**Key Steps in Plasma Activation:**

1. **Vacuum Chamber Setup:**

   The material to be treated is placed inside a vacuum chamber or reactor equipped with electrodes and gas inlet ports.

2. **Gas Introduction:**

   A suitable process gas, such as oxygen, nitrogen, argon, or hydrogen, is introduced into the chamber at a controlled flow rate.

3. **Plasma Ignition:**

   A radiofrequency (RF) or microwave power source is applied to the electrodes, creating an electrical discharge that ionizes the process gas and generates plasma.

4. **Surface Treatment:**

   The material surface is exposed to the plasma for a defined duration, during which energetic plasma species interact with the surface, leading to surface cleaning, activation, etching, or functionalization.

5. **Gas Purging:**

   After the treatment, the chamber may be purged with an inert gas or evacuated to remove residual gases and by-products before opening.

**Applications of Plasma Activation in Cleanroom Environments:**

Plasma activation finds diverse applications in semiconductor fabrication, microelectronics, MEMS (microelectromechanical systems), NEMS (nanoelectromechanical systems), and other microfabrication processes. Some common applications include:

1. **Surface Cleaning:**

   Plasma activation effectively removes organic contaminants, residues, and native oxide layers from material surfaces, ensuring clean and reactive surfaces for subsequent processing steps such as deposition, bonding, and lithography.

2. **Surface Activation:**

   Plasma treatment enhances surface wettability, adhesion, and bonding properties by introducing polar functional groups and surface radicals, promoting strong interfacial interactions between materials and adhesives, coatings, or deposited layers.

3. **Surface Functionalization:**

   Plasma activation facilitates the grafting or deposition of functional groups, nanoparticles, or thin films onto material surfaces, enabling tailored surface properties such as hydrophilicity, hydrophobicity, biocompatibility, or antimicrobial activity.

4. **Photoresist Adhesion:**

   Plasma activation improves the adhesion of photoresist materials to substrates by modifying the surface energy and chemistry of the substrate, ensuring uniform coating and pattern transfer during photolithography processes.

5. **Surface Modification:** Plasma treatment can selectively modify surface chemistry, roughness, and morphology of materials, allowing precise control over surface properties for specific applications such as surface passivation, biofunctionalization, or sensor development.

**Significance in Cleanroom Environments:**

In cleanroom environments dedicated to semiconductor manufacturing and microfabrication, plasma activation plays a critical role in ensuring the cleanliness, uniformity, and reliability of surface treatments. By providing controlled and repeatable surface modifications without introducing contaminants or residues, plasma activation contributes to the production of high-quality microdevices, integrated circuits, sensors, and other advanced materials with superior performance and functionality. Additionally, plasma activation enables process flexibility, scalability, and compatibility with a wide range of substrates and materials, making it an indispensable tool for achieving precise and reliable surface modifications in cleanroom environments.

**Conclusion:**

Plasma activation is a versatile and effective surface modification technique that finds widespread use in cleanroom environments for enhancing the surface properties of materials in semiconductor fabrication and microfabrication processes. By providing controlled and uniform surface treatments without compromising cleanliness or reliability, plasma activation enables the production of advanced microdevices, sensors, and integrated circuits with tailored surface functionalities and improved performance. Understanding the principles and applications of plasma activation is essential for optimizing surface treatment processes and achieving desired material properties in cleanroom environments dedicated to semiconductor manufacturing and microfabrication.

## 20.9 Critical Drier:
### Ensuring Optimal Drying in Cleanroom Environments.

The critical drier, also known as a spin rinse dryer (SRD) or megasonic dryer, is a vital component in cleanroom environments, particularly in semiconductor fabrication facilities, where moisture control is paramount. This subchapter delves into the operating principles of critical driers, their applications in cleanroom processes, and their significance in achieving precise and efficient drying of substrates and wafers.

**Operating Principle:**

Critical driers utilize a combination of centrifugal force, megasonic energy, and controlled temperature to remove residual water and contaminants from the surface of substrates, wafers, or other materials. The drying process typically involves the following steps:

1. **Spin Drying:**

   The wet substrate is placed onto a spinning chuck or platform within the drier chamber. As the chuck spins at high speed, centrifugal force drives the bulk of the liquid away from the substrate's surface, reducing the risk of water droplets or films.

2. **Megasonic Treatment:**

   In conjunction with spin drying, megasonic transducers emit high-frequency acoustic waves (typically in the MHz range) into the liquid film remaining on the substrate. These waves create cavitation bubbles that implode near the surface, generating localized shockwaves and microstreaming, which effectively dislodge and remove trapped particles, residues, and dissolved gases.

3. **Rinsing and Purging:**

   Depending on the specific application and cleanliness requirements, the substrate may undergo additional rinsing with deionized water or other solvents to further remove residual contaminants. The drier chamber may then be purged with clean, dry nitrogen or filtered air to assist in the evaporation of remaining moisture and ensure a dry, particle-free environment.

**Applications of Critical Driers in Cleanroom Environments:**

Critical driers play a crucial role in various semiconductor manufacturing processes and cleanroom operations, including:

1. **Wafer Drying:**

   After wet processing steps such as cleaning, etching, or chemical deposition, semiconductor wafers must be thoroughly dried to prevent water spots, surface defects, or contamination that could compromise device performance or yield. Critical driers ensure rapid and uniform drying of wafers without the risk of watermarks or residues.

2. **Photoresist Stripping:**

   Following photolithography processes, substrates may undergo photoresist stripping to remove patterned photoresist layers. Critical driers assist in the drying of substrates after resist stripping baths, ensuring complete removal of residual solvent and contaminants without damaging underlying layers or structures.

3. **Surface Preparation:**

   Critical driers are employed in surface preparation steps such as substrate cleaning, surface activation, or bonding, where the presence of moisture or contaminants can interfere with subsequent processing steps. By providing controlled and efficient drying, critical driers facilitate the preparation of clean, dry surfaces for further processing or analysis.

4. **MEMS/NEMS Fabrication:**

   In microelectromechanical systems (MEMS) and nanoelectromechanical systems (NEMS) fabrication, critical driers are utilized to dry delicate microstructures, sensors, or actuators without causing mechanical damage or stiction. The gentle drying process preserves the integrity and functionality of MEMS/NEMS devices while ensuring high reliability and performance.

**Significance in Cleanroom Environments:**

Critical driers play a critical role in maintaining cleanliness, quality, and reliability in cleanroom environments by providing rapid, efficient, and particle-free drying of substrates and wafers. By eliminating residual water and contaminants, critical driers contribute to the prevention of defects, yield loss, and reliability issues in semiconductor devices and microelectronics. Additionally, the gentle and non-contact drying process of critical driers minimizes the risk of substrate damage or contamination, making them indispensable tools for achieving precise and reliable processing in cleanroom environments.

**Conclusion:**

The critical drier is an essential tool in cleanroom environments, offering rapid, efficient, and particle-free drying of substrates and wafers in semiconductor fabrication and microfabrication processes. By combining spin drying with megasonic treatment, critical driers ensure thorough removal of residual water, contaminants, and dissolved gases from surfaces, thereby preventing defects, yield loss, and reliability issues in semiconductor devices and microelectronics. Understanding the principles and applications of critical driers is essential for optimizing drying processes and achieving desired cleanliness and quality standards in cleanroom environments dedicated to semiconductor manufacturing and microfabrication.

## 20.10 Wet Benches:

**Essential Tools for Wet Processing in Cleanroom Environments.**

Wet benches are foundational tools in cleanroom environments, serving as specialized workstations designed for wet chemical processing and handling of substrates, wafers, and other sensitive materials. This subchapter explores the types of wet benches, their applications in cleanroom operations, and the significance of stainless steel construction in certain configurations.

**Types of Wet Benches:**

1. **Manual Wet Benches:**

   These traditional wet benches feature manually operated sinks, chemical dispensers, and rinse stations. Operators perform wet processing steps such as cleaning, etching, or coating by manually manipulating substrates and controlling chemical flow rates. Manual wet benches offer versatility and flexibility for research, prototyping, and low-volume production applications.

2. **Semi-Automated Wet Benches:**

   Semi-automated wet benches incorporate automated features such as programmable recipes, robotic wafer handling, and automated chemical delivery systems. These benches streamline wet processing workflows, improve process repeatability and consistency, and reduce operator error and contamination risks. Semi-automated wet benches are commonly used in medium to high-volume production environments where precision, throughput, and yield are paramount.

3. **Fully Automated Wet Benches:**

   Fully automated wet benches feature advanced robotics, process control, and integrated metrology systems for fully autonomous wet processing operations. These benches offer unparalleled process control, throughput, and yield optimization capabilities, making them ideal for high-volume manufacturing of semiconductor devices, MEMS/NEMS devices, and photonic components. Fully automated wet benches are integral components of advanced fabrication facilities (fabs) and semiconductor foundries.

**Applications of Wet Benches in Cleanroom Environments:**

Wet benches play a pivotal role in a wide range of wet chemical processing applications in cleanroom environments, including:

1. **Cleaning and Surface Preparation:**

   Wet benches are used for substrate cleaning, surface activation, and preparation steps prior to deposition, lithography, or bonding processes. Wet cleaning techniques may involve solvent cleaning, acid etching, or plasma treatment to remove contaminants, residues, or native oxides from substrate surfaces.

2. **Etching and Stripping:**

   Wet benches facilitate wet etching processes for selective removal of materials from substrate surfaces. Wet etching techniques include isotropic and anisotropic etching of silicon, silicon dioxide, metals, and other materials using etchants such as acids, bases, or specialized chemical solutions. Wet benches also support resist stripping processes to remove patterned photoresist layers after lithography steps.

3. **Surface Modification and Functionalization:**

   Wet benches enable surface modification and functionalization processes to tailor surface properties for specific applications. Wet chemical treatments may involve surface passivation, oxidation, or deposition of functional coatings to enhance adhesion, wetting, or biocompatibility of substrate surfaces.

4. **Wafer Cleaning and Rinsing:**

   Wet benches provide dedicated stations for wafer cleaning and rinsing using deionized water or solvent-based rinses. Cleanroom operators can perform multiple rinse steps to ensure thorough removal of residual chemicals and contaminants from substrates before proceeding to subsequent processing steps.

**Significance of Stainless Steel Construction:**

Many wet benches feature stainless steel construction for their work surfaces, sink basins, and chemical containment components. Stainless steel offers several advantages in cleanroom wet processing environments, including:

1. **Chemical Resistance:**

   Stainless steel is highly resistant to corrosion, oxidation, and chemical attack from a wide range of acids, bases, solvents, and process chemicals used in wet processing operations. This resistance ensures long-term durability and integrity of wet bench components, minimizing contamination risks and maintenance requirements.

2. **Cleanliness and Hygiene:**
Stainless steel surfaces are smooth, non-porous, and easy to clean, making them ideal for cleanroom environments where cleanliness, sterility, and contamination control are paramount. Stainless steel resists microbial growth, staining, and particulate accumulation, ensuring compliance with stringent cleanliness standards and regulatory requirements.

3. **Strength and Stability:**
Stainless steel exhibits excellent mechanical properties, including high strength, rigidity, and dimensional stability, which are critical for supporting heavy loads, equipment, and substrates in wet processing applications. Stainless steel wet benches provide a robust and stable platform for performing precision wet processing steps without compromising accuracy or reliability.

**Conclusion:**
Wet benches are indispensable tools in cleanroom environments, providing essential infrastructure for wet chemical processing, cleaning, and surface preparation of substrates and wafers in semiconductor fabrication, MEMS/NEMS manufacturing, and photonic device production. Understanding the types of wet benches, their applications, and the significance of stainless steel construction is essential for optimizing wet processing workflows, ensuring process integrity, and maintaining cleanliness and reliability in cleanroom environments dedicated to advanced microfabrication and nanofabrication technologies.

## 20.11 Electronic Beam Lithography EBL vs. Regular Lithography: A Comparative Analysis

Electronic Beam Lithography (EBL) and regular lithography are two fundamental techniques employed in semiconductor manufacturing and nanofabrication for pattern transfer onto substrates. This subchapter provides a comparative analysis of EBL and regular lithography, highlighting their respective principles, advantages, limitations, and applications in cleanroom environments.

**Principles of Operation:**

1. **Electronic Beam Lithography (EBL):**
EBL utilizes a focused beam of electrons to write patterns directly onto a substrate coated with a resist material. The electron beam is controlled by electromagnetic lenses and deflection systems, enabling precise manipulation

of the beam position and intensity. EBL systems typically operate in a vacuum environment to minimize electron scattering and achieve high-resolution patterning at the nanoscale.

2. **Regular Lithography:**

Regular lithography, also known as optical or photolithography, relies on optical projection systems to transfer patterns from photomasks onto photoresist-coated substrates. Ultraviolet (UV) light is used to expose the photoresist through the photomask, creating a latent image that is subsequently developed to define the desired pattern. Regular lithography techniques include contact, proximity, and projection lithography, each offering different resolution capabilities and process complexities.

**Advantages of EBL:**

1. **High Resolution:**

EBL systems can achieve extremely high resolution patterning down to the sub-10 nanometer scale, making them ideal for nanotechnology research, semiconductor device prototyping, and advanced lithographic applications requiring fine features and intricate geometries.

2. **Direct Write Capability:**

EBL enables direct writing of patterns without the need for photomasks, facilitating rapid prototyping, design iterations, and customization of patterns on-demand. This direct-write capability enhances flexibility and versatility in pattern generation, reducing turnaround times and cost overhead associated with mask fabrication.

3. **Pattern Complexity:**

EBL systems offer unparalleled flexibility in patterning complex, non-periodic, and irregular geometries with high fidelity and accuracy. This capability is particularly advantageous for applications requiring customized micro and nanostructures, such as photonic devices, microfluidic systems, and bioMEMS devices.

**Advantages of Regular Lithography:**

1. **Throughput:**

Regular lithography techniques, especially proximity and projection lithography, offer high throughput capabilities, enabling rapid and cost-effective replication of patterns across large substrate areas. This high throughput is essential for mass production of semiconductor devices, integrated circuits, and microelectronic components in commercial fabrication facilities (fabs).

2. **Process Compatibility:**
Regular lithography processes are compatible with a wide range of photoresist materials, substrates, and processing conditions, making them suitable for diverse applications in semiconductor manufacturing, MEMS fabrication, and microelectronics packaging. This compatibility facilitates integration with existing process flows and equipment in cleanroom environments.

3. **Cost-Effectiveness:**
Regular lithography techniques benefit from mature infrastructure, standardized processes, and economies of scale, resulting in lower overall manufacturing costs compared to EBL for high-volume production applications. The availability of off-the-shelf photomasks and optical exposure systems further contributes to cost-effectiveness and process scalability.

**Limitations and Challenges:**

1. **Resolution Limitations:**
Regular lithography techniques are limited by diffraction effects, optical aberrations, and photomask resolution, constraining the achievable feature sizes and pattern fidelity. EBL, on the other hand, can overcome these limitations and achieve sub-diffraction-limited resolution due to the shorter wavelength of electrons.

2. **Complexity and Cost:**
EBL systems are inherently more complex and expensive than regular lithography equipment, requiring specialized infrastructure, vacuum chambers, electron sources, and precise beam control mechanisms. The initial capital investment and maintenance costs associated with EBL may be prohibitive for some research laboratories and small-scale fabrication facilities.

**Applications and Future Trends:**

1. **EBL Applications:**
EBL finds applications in research fields such as nanotechnology, quantum computing, photonics, and plasmonics, where ultra-high resolution patterning and precise control over nanostructures are essential. EBL is also used for maskless lithography, direct writing of electron-beam resist patterns, and template fabrication for nanoimprint lithography.

2. **Regular Lithography Applications:**
Regular lithography techniques remain indispensable for high-volume manufacturing of integrated circuits, microprocessors, memory devices, and sensor arrays in the semiconductor industry. Continuous advancements in optical lithography, including immersion lithography, multi-patterning, and computational lithography, enable further scaling of feature sizes and increased device densities.

**Conclusion:**

Electronic Beam Lithography (EBL) and regular lithography are complementary techniques with distinct advantages and limitations in cleanroom environments. While EBL offers unparalleled resolution, direct write capability, and flexibility for research and low-volume prototyping, regular lithography techniques excel in high-throughput manufacturing of semiconductor devices and microelectronic components. Understanding the differences between EBL and regular lithography is essential for selecting the most appropriate lithographic technique based on the specific requirements of the application, desired feature sizes, throughput considerations, and available resources in cleanroom facilities.

## 20.12 Furnace Types and Their Applications in Cleanroom Environments

Furnaces are essential equipment in cleanroom environments for thermal processing of semiconductor wafers and other substrates. This subchapter provides an overview of furnace types commonly used in cleanrooms, along with their respective operating principles, applications, and advantages.

**1. Horizontal Tube Furnaces:**

- **Operating Principle:**

Horizontal tube furnaces consist of a cylindrical chamber with a horizontally positioned quartz or ceramic tube. Heating elements surround the tube, providing uniform heating along its length. Substrates are loaded into the tube and heated to high temperatures under controlled atmospheric conditions.

- **Applications:**

Horizontal tube furnaces are versatile tools used for various high-temperature processes, including thermal oxidation, diffusion, annealing, and chemical vapor deposition (CVD). They are suitable for processing semiconductor wafers, thin film deposition, and heat treatment of materials in research and manufacturing applications.

- **Advantages:**

Horizontal tube furnaces offer excellent temperature uniformity, precise temperature control, and scalability for batch processing of multiple substrates. They are compatible with inert gas atmospheres and vacuum conditions, enabling a wide range of thermal processing techniques.

**2. Vertical Tube Furnaces:**

- **Operating Principle:**

Vertical tube furnaces feature a vertical orientation, with heating elements surrounding a quartz or ceramic tube placed vertically. Substrates are loaded into the top of the tube and heated as they descend through the furnace chamber. Heating zones along the tube length allow for temperature gradients and controlled thermal profiles.

- **Applications:**

Vertical tube furnaces are commonly used for processes such as epitaxial growth, diffusion, annealing, and crystal growth. They are particularly suitable for applications requiring precise control of temperature gradients and thermal cycling, such as semiconductor device fabrication and material synthesis.

- **Advantages:**

Vertical tube furnaces offer efficient heat transfer, uniform temperature distribution, and programmable heating profiles. They are well-suited for processing large-area substrates, tall structures, and multi-layered films. Vertical orientation facilitates easy loading and unloading of substrates, minimizing handling-induced contamination.

**3. Rapid Thermal Processing (RTP) Systems:**

- **Operating Principle:**

RTP systems utilize intense radiant heating to achieve rapid heating and cooling cycles for semiconductor processing. Substrates are heated by high-intensity lamps or lasers for short durations, followed by rapid quenching to ambient temperature. RTP systems provide precise temperature control and rapid thermal response.

- **Applications:**

RTP systems are employed for rapid thermal annealing, dopant activation, oxide growth, and silicide formation in semiconductor manufacturing. They are suitable for processes requiring ultra-fast heating and cooling rates, reduced thermal budget, and precise temperature control at the nanosecond timescale.

- **Advantages:**

RTP systems offer significantly reduced processing times compared to conventional furnace techniques, enabling higher throughput and energy efficiency. They minimize thermal budget, reducing diffusion and interfacial reactions, while achieving superior film quality and device performance. RTP systems are ideal for advanced CMOS technology nodes and emerging semiconductor applications.

**Conclusion:**

Furnaces play a crucial role in cleanroom environments for thermal processing of semiconductor wafers and thin film deposition. Horizontal tube furnaces, vertical tube furnaces, and rapid thermal processing (RTP) systems offer distinct advantages and capabilities for various applications in semiconductor manufacturing, research, and development. Understanding the operating principles, applications, and advantages of different furnace types is essential for optimizing thermal processing techniques, improving device performance, and advancing semiconductor technology.

## 20.13 Photo Resists

Photoresists are critical materials used in semiconductor manufacturing and microfabrication processes within cleanroom environments. This subchapter provides an overview of photoresists, their types, properties, and applications in cleanroom technology.

**1. Introduction to Photoresists:**

- **Definition:**
Photoresists are light-sensitive materials that undergo chemical changes when exposed to ultraviolet (UV) or visible light. They are used to transfer patterns onto semiconductor wafers or substrates during photolithography processes.

- **Types of Photoresists:**
  - **Positive Photoresists:**
  Positive photoresists become more soluble upon exposure to light. Exposed regions are removed during development, leaving behind the desired pattern.
  - **Negative Photoresists:**
  Negative photoresists become less soluble upon exposure to light. Exposed regions polymerize or crosslink, forming the desired pattern upon development.

**2. Properties of Photoresists:**

- **Photosensitivity:**
Photoresists exhibit varying degrees of sensitivity to different wavelengths of light, depending on their chemical composition and formulation.

- **Resolution:**
The ability of a photoresist to reproduce fine features and patterns is crucial for high-resolution lithography processes.

- **Adhesion:**
Photoresists must adhere well to substrates to ensure pattern transfer and dimensional fidelity during subsequent processing steps.

- **Chemical Compatibility:**
Photoresists should be compatible with developer solutions, etchants, and other chemicals used in the fabrication process.

**3. Applications of Photoresists in Cleanroom Technology:**

- **Photolithography:**
Photoresists are used in photolithography processes to transfer mask patterns onto semiconductor wafers or substrates. This technique is fundamental to

patterning features at the microscale and nanoscale in semiconductor device fabrication.

- **Etching Masks:**
Patterned photoresist layers serve as masks for etching processes, where exposed regions are selectively removed to define device features, interconnects, and circuitry.
- **Implantation Masks:**
In ion implantation processes, photoresist masks are used to define doping profiles and implantation areas on semiconductor substrates.
- **Lift-off Processes:**
Photoresists are employed in lift-off processes to define and lift deposited metal layers, creating fine structures and features with high aspect ratios.
- **Template for Nanoimprint Lithography:**
Photoresists can serve as templates for nanoimprint lithography, enabling the replication of nanoscale patterns onto substrates for various applications in nanotechnology and nanofabrication.

**Conclusion:**
Photoresists are indispensable materials in cleanroom environments for semiconductor manufacturing, microfabrication, and nanotechnology applications. Their unique properties and versatile applications make them essential for defining patterns, creating device features, and achieving high-resolution patterning on semiconductor wafers and substrates. Understanding the properties and applications of photoresists is crucial for optimizing lithography processes, enhancing device performance, and advancing cleanroom technology.

## 20.14 Chemical Storage Cabinets

Chemical storage cabinets play a crucial role in maintaining a safe and organized cleanroom environment, particularly in facilities where hazardous chemicals are used in semiconductor manufacturing, microfabrication, and nanotechnology processes. This subchapter provides an overview of chemical storage cabinets, their importance, and the necessity of connecting them to the exhaust system for proper ventilation and safety.

**1. Introduction to Chemical Storage Cabinets:**
- **Purpose:**

Chemical storage cabinets are specially designed containers used to store hazardous chemicals, acids, solvents, and reagents safely within cleanroom facilities.

- **Safety Features:**
  These cabinets are constructed with materials resistant to chemical corrosion and fire, providing a secure storage environment to minimize the risk of chemical spills, leaks, and accidents.
- **Organization:** Chemical storage cabinets are equipped with shelves, compartments, and labeling systems to facilitate the organized storage and easy retrieval of chemicals, ensuring efficient workflow and inventory management.

## 2. Importance of Chemical Storage Cabinets in Cleanrooms:

- **Hazard Mitigation:**
  Proper storage of hazardous chemicals in designated cabinets helps mitigate the risk of chemical exposure, contamination, and potential health hazards to cleanroom personnel.
- **Compliance:**
  Compliance with regulatory standards and safety guidelines is essential in cleanroom environments. Chemical storage cabinets ensure adherence to safety protocols and regulatory requirements governing the handling and storage of hazardous substances.
- **Prevention of Cross-Contamination:**
  Segregating chemicals based on their compatibility and hazard level prevents cross-contamination and chemical reactions, safeguarding the integrity of cleanroom processes and sensitive materials.
- **Emergency Preparedness:**
  Chemical storage cabinets are equipped with spill containment trays, spill kits, and emergency response supplies to facilitate prompt and effective mitigation of chemical spills, leaks, or accidents.

## 3. Connection to Exhaust System:

- **Ventilation Requirements:**
  To maintain a safe working environment and prevent the accumulation of hazardous vapors, chemical storage cabinets must be connected to the cleanroom's exhaust system.
- **Fume Extraction:**
  Connecting chemical storage cabinets to the exhaust system ensures the efficient removal of volatile vapors, fumes, and airborne contaminants generated during chemical handling and storage.
- **Air Quality Control:**

Continuous ventilation and airflow through the cabinets help maintain air quality standards within the cleanroom, reducing the risk of exposure to hazardous chemicals and ensuring the well-being of cleanroom personnel.

- **Regulatory Compliance:**

Proper ventilation and connection to the exhaust system are essential for compliance with regulatory requirements and safety standards governing the storage and handling of hazardous chemicals in cleanroom environments.

**Conclusion:**

Chemical storage cabinets are indispensable safety fixtures in cleanroom facilities, providing secure storage solutions for hazardous chemicals while minimizing the risk of accidents, contamination, and exposure. Connecting these cabinets to the exhaust system ensures proper ventilation, fume extraction, and compliance with regulatory standards, enhancing overall safety and environmental control in cleanroom environments.

## 20.15 DI water Stations

Deionized (DI) water stations are essential components of cleanroom facilities, serving as critical utilities for various manufacturing, fabrication, and research processes. This subchapter provides an overview of DI water stations, their significance, and applications within cleanroom environments.

**1. Introduction to DI Water Stations:**

- **Purpose:**

DI water stations are specialized systems designed to produce and dispense deionized water, also known as ultrapure water, for use in cleanroom operations.

- **Water Purification:**

These stations employ advanced filtration and purification techniques to remove impurities, ions, dissolved solids, and contaminants from tap or feed water, resulting in high-purity DI water suitable for critical applications.

- **Storage and Dispensing:**

DI water stations typically feature storage tanks or reservoirs to store purified water, along with dispensing mechanisms such as faucets, hoses, or nozzles for convenient access and distribution of DI water to cleanroom users and processes.

**2. Importance of DI Water in Cleanroom Processes:**

- **Critical Utility:**

DI water serves as a universal solvent and cleaning agent in cleanroom environments, supporting a wide range of applications, including wafer rinsing, substrate cleaning, chemical dilution, and laboratory experiments.

- **Process Compatibility:**
  The ultra-high purity of DI water makes it ideal for use in semiconductor manufacturing, microelectronics fabrication, nanotechnology research, and precision optics industries, where even trace contaminants can compromise product quality and performance.
- **Surface Preparation:**
  DI water is commonly used for surface cleaning, rinsing, and preparation of substrates, wafers, and components before subsequent processing steps such as lithography, etching, deposition, and bonding, ensuring optimal adhesion, adhesion, and performance.
- **Analytical Applications:**
  In addition to manufacturing processes, DI water is utilized in analytical laboratories, quality control testing, and metrology applications for preparing standard solutions, calibration standards, and instrument rinsing to ensure accurate and reproducible results.

### 3. Features and Components of DI Water Stations:

- **Water Purification System:**
  DI water stations incorporate multi-stage purification systems comprising pre-filters, activated carbon filters, reverse osmosis membranes, and ion exchange resins to remove particulates, organics, dissolved salts, and ions from feed water, producing ultrapure DI water with low conductivity and high resistivity.
- **Storage and Distribution:**
  The stations are equipped with storage tanks or reservoirs made of corrosion-resistant materials such as stainless steel or polypropylene to maintain the purity and integrity of DI water. Dispensing mechanisms such as faucets, valves, or automated pumps allow controlled delivery of DI water to cleanroom users and processes.
- **Monitoring and Control:**
  Advanced DI water stations feature monitoring and control systems to continuously monitor water quality parameters such as conductivity, resistivity, pH, and total organic carbon (TOC), ensuring compliance with purity standards and regulatory requirements. Alarms and alerts notify users of any deviations or abnormalities in water quality.

**Conclusion:**

DI water stations play a vital role in cleanroom operations, providing ultrapure water for critical manufacturing, fabrication, and research processes. By delivering high-purity water with minimal contaminants, these stations ensure the integrity, reliability, and performance of cleanroom processes across various industries and applications, contributing to product quality, yield, and innovation.

## 20.16 Liquid Nitrogen Dewars

Liquid nitrogen dewars are indispensable tools in cleanroom facilities, serving a crucial role in various processes, including semiconductor manufacturing, microfabrication, and nanotechnology research. This subchapter provides a detailed overview of liquid nitrogen dewars, their significance, and applications within cleanroom environments, particularly in connection with deep etchers utilizing the Bosch Process.

**1. Introduction to Liquid Nitrogen Dewars:**

- **Purpose:**

Liquid nitrogen dewars are specialized containers designed to store and dispense liquid nitrogen, a cryogenic fluid with a boiling point of -196°C (-321°F), in a controlled and safe manner.

- **Cryogenic Cooling:**

Liquid nitrogen is commonly utilized in cleanroom environments for cryogenic cooling applications, including temperature control, thermal management, and cryopreservation of samples, materials, and equipment.

- **Safe Handling:**

Dewars are engineered with robust construction materials, such as stainless steel or aluminum, and equipped with safety features such as pressure relief valves, vacuum insulation, and sturdy handles for safe handling and transportation of cryogenic fluids.

**2. Importance of Liquid Nitrogen in Cleanroom Processes:**

- **Cryogenic Etching:**

Liquid nitrogen plays a vital role in cryogenic etching processes, particularly in deep etching techniques such as the Bosch Process, utilized for the fabrication of high-aspect-ratio microstructures, MEMS devices, and sensors in semiconductor and MEMS industries.

- **Bosch Process:**

In the Bosch Process, liquid nitrogen is used as a coolant and cryogen to maintain low temperatures within the process chamber during the etching cycles, enabling precise control over etch rates, sidewall profiles, and feature dimensions.

- **Temperature Control:**

By introducing liquid nitrogen into the process chamber, the temperature of the substrate and etch byproducts can be reduced significantly, minimizing thermal effects, sidewall redeposition, and polymer accumulation, resulting in improved etch selectivity and pattern fidelity.

- **Etch Rate Enhancement:**

Cryogenic cooling enhances the etch rate of certain materials, such as silicon (Si), silicon dioxide (SiO2), and silicon nitride (Si3N4), by promoting chemical reactions and physical processes at lower temperatures, leading to faster and more efficient material removal.

**3. Integration with Deep Etchers and Bosch Process:**

- **Etcher Cooling System:**

Liquid nitrogen dewars are integrated into the cooling systems of deep etchers, providing a continuous supply of cryogenic coolant to maintain optimal process temperatures during etching operations.

- **Temperature Control:**

In the Bosch Process, liquid nitrogen dewars are connected to the process chamber or cooling jacket of the etcher to cool the silicon substrate and etch byproducts, preventing overheating and ensuring uniform etching across the wafer surface.

- **Process Stability:**

By controlling the temperature with liquid nitrogen, the Bosch Process achieves superior process stability, repeatability, and control over etch profiles, minimizing defects, roughness, and variation in feature dimensions.

**Conclusion:**

Liquid nitrogen dewars are essential components of cleanroom facilities, providing cryogenic cooling for various processes, including deep etching in the Bosch Process. By supplying controlled temperatures to process chambers, these dewars enable precise control over etch rates, sidewall profiles, and feature dimensions, contributing to the fabrication of advanced microstructures and devices in semiconductor and MEMS industries. Their integration with deep etchers ensures reliable and efficient operation, enhancing process stability, repeatability, and yield in cleanroom environments.

## 20.17 Exhaust Gas Scrubbers

Exhaust gas scrubbers play a critical role in maintaining air quality and safety within cleanroom facilities by effectively removing hazardous gases and contaminants from exhaust streams. This subchapter explores the importance of exhaust gas scrubbers, their applications in cleanroom environments, and strategies to prevent clogging for optimal performance.

**1. Introduction to Exhaust Gas Scrubbers:**

- **Purpose:**

Exhaust gas scrubbers, also known as gas abatement systems, are designed to capture, neutralize, and remove hazardous gases and volatile organic compounds (VOCs) emitted during semiconductor manufacturing, chemical processing, and other industrial processes.

- **Gas Abatement:**

Scrubbers utilize various chemical and physical processes, such as adsorption, absorption, oxidation, and chemical reaction, to effectively neutralize and eliminate toxic and harmful gases from exhaust streams, ensuring compliance with environmental regulations and safety standards.

- **Air Quality:**

By removing hazardous gases, scrubbers help maintain cleanroom air quality, prevent contamination of sensitive equipment and processes, and protect personnel from exposure to harmful emissions, thus ensuring a safe and healthy working environment.

## 2. Applications of Exhaust Gas Scrubbers in Cleanrooms:

- **Chemical Etching:**

In semiconductor fabrication and microfabrication processes, exhaust gas scrubbers are employed to remove byproducts and waste gases generated during wet chemical etching, such as hydrochloric acid (HCl), sulfuric acid ($H_2SO_4$), and hydrogen fluoride (HF), to prevent corrosion and contamination of equipment.

- **Thin Film Deposition:**

During chemical vapor deposition (CVD) and physical vapor deposition (PVD) processes, scrubbers capture and neutralize precursor gases and reaction byproducts, such as silane ($SiH_4$), tungsten hexafluoride ($WF_6$), and nitrogen trifluoride ($NF_3$), to prevent chamber fouling and maintain process integrity.

- **Etch and Strip Processes:**

Scrubbers remove toxic and corrosive gases, such as chlorine ($Cl_2$), bromine ($Br_2$), and sulfur hexafluoride ($SF_6$), released during dry etching and plasma stripping processes, minimizing equipment degradation and ensuring consistent device performance.

## 3. Preventing Clogging in Exhaust Gas Scrubbers:

- **Regular Maintenance:**

Routine inspection, cleaning, and maintenance of exhaust gas scrubbers are essential to prevent clogging and ensure uninterrupted operation. Maintenance tasks include checking filter media, replacing absorbent materials, and cleaning internal components to remove accumulated residues and contaminants.

- **Optimized Operation:** Proper adjustment of gas flow rates, temperature, and scrubbing parameters helps optimize scrubber performance and prevent clogging by ensuring efficient gas-liquid contact and absorption of contaminants.
- **Monitoring Systems:** Continuous monitoring of exhaust gas flow rates, pressure differentials, and scrubber performance parameters allows early detection of potential issues, such as reduced gas removal efficiency or increased pressure drop, enabling timely intervention and preventive maintenance to avoid clogging.
- **Chemical Treatment:** Periodic treatment of scrubber media with cleaning agents or regenerants helps dissolve and remove deposited solids, scale, and fouling compounds, preventing clogging and restoring scrubber performance to optimal levels.

**Conclusion:**

Exhaust gas scrubbers play a vital role in maintaining air quality and safety in cleanroom environments by effectively removing hazardous gases and contaminants from exhaust streams. By capturing and neutralizing toxic emissions, scrubbers prevent equipment corrosion, process contamination, and personnel exposure, ensuring a safe and healthy working environment. Preventive maintenance, optimized operation, and regular monitoring are essential strategies to prevent clogging and maintain scrubber performance for reliable and efficient gas abatement in cleanroom facilities.

## 20.18 Exhaust Systems

**Exhaust Systems in Cleanroom Environments**

Exhaust systems are fundamental components of cleanroom facilities, serving to remove contaminants, maintain air quality, and ensure a safe and controlled working environment. This subchapter delves into the key aspects of exhaust systems, their significance in cleanrooms, and considerations for effective operation and maintenance.

**1. Importance of Exhaust Systems:**

- **Contaminant Removal:**

Exhaust systems play a crucial role in eliminating airborne particles, chemical fumes, and volatile organic compounds (VOCs) generated during manufacturing processes, preventing contamination of cleanroom environments and sensitive equipment.

- **Air Quality Control:**

By expelling pollutants and maintaining proper ventilation rates, exhaust systems help control indoor air quality, regulate temperature and humidity levels, and minimize the risk of exposure to hazardous substances, ensuring a safe and comfortable working environment for cleanroom personnel.

- **Regulatory Compliance:**
Cleanroom exhaust systems are designed and operated in compliance with regulatory standards and industry guidelines to meet stringent cleanliness and safety requirements, safeguarding product integrity and personnel health.

## 2. Components of Exhaust Systems:

- **Exhaust Fans:**
Centrifugal or axial fans are employed to create negative pressure within cleanrooms, drawing contaminated air and gases out of the facility and expelling them into the atmosphere or through filtration systems.

- **Ductwork:**
Ductwork networks, comprising ducts, dampers, and vents, distribute exhaust air from process areas to exterior discharge points or filtration units, ensuring efficient removal of contaminants while minimizing cross-contamination and airflow disruptions.

- **Filtration Systems:**
HEPA (High-Efficiency Particulate Air) filters or activated carbon filters may be integrated into exhaust systems to capture and remove airborne particles, allergens, and chemical odors, enhancing air purification and environmental protection.

- **Exhaust Stacks:**
Vertical exhaust stacks or chimneys facilitate the safe discharge of contaminated air into the atmosphere, preventing re-entry of pollutants into the cleanroom environment and minimizing environmental impact through dispersion and dilution.

## 3. Considerations for Exhaust System Design and Operation:

- **Airflow Management:**
Proper design and layout of exhaust systems, including the placement and sizing of exhaust fans, ducts, and outlets, are critical to ensure uniform airflow distribution, efficient contaminant capture, and optimal ventilation performance throughout the cleanroom facility.

- **Controlled Ventilation:**
Automated control systems, equipped with sensors and actuators, regulate exhaust fan operation, airflow rates, and pressure differentials based on real-time monitoring of airborne particle levels, temperature, humidity, and

chemical concentrations, maintaining optimal environmental conditions and energy efficiency.

- **Noise and Vibration Control:**
Implementation of soundproofing materials, vibration isolation mounts, and noise-reducing baffles helps minimize noise and vibration generated by exhaust fans and ductwork, ensuring a quiet and comfortable working environment for cleanroom personnel.

- **Maintenance and Monitoring:**
Regular inspection, cleaning, and maintenance of exhaust systems are essential to prevent airway obstructions, filter clogging, and fan malfunctions, ensuring continuous operation and compliance with cleanliness and safety standards.

**Conclusion:**

Exhaust systems are integral components of cleanroom facilities, providing essential ventilation, contaminant removal, and environmental control to safeguard product quality, process integrity, and personnel health. Through proper design, operation, and maintenance, exhaust systems contribute to the effective management of indoor air quality, regulatory compliance, and overall cleanliness in cleanroom environments, supporting the success and sustainability of high-tech manufacturing and research activities.

## 20.19 Gas lines Thermal jackets

Gas line thermal jackets play a crucial role in semiconductor manufacturing facilities by preventing the condensation of gases used in various processes. This subchapter explores the importance of thermal jackets, their application in cleanroom environments, and the gases that may require insulation to maintain optimal temperature conditions.

### 1. Importance of Gas Line Thermal Jackets:

- **Condensation Prevention:**
In semiconductor manufacturing, certain process gases are stored and transported at high pressures and temperatures. When these gases encounter lower temperatures in ambient conditions or within gas delivery lines, they can condense, leading to undesirable outcomes such as reduced process efficiency, equipment corrosion, and contamination risks.

- **Temperature Control:**
Gas line thermal jackets provide insulation and heat retention properties, maintaining the temperature of process gases within specified ranges during storage, distribution, and delivery. By preventing temperature fluctuations and

condensation formation, thermal jackets ensure consistent gas flow rates and chemical reactions, optimizing process performance and product quality.

- **Equipment Protection:**
Thermal jackets protect gas delivery systems, valves, regulators, and associated components from thermal stress, moisture ingress, and corrosion caused by condensation or exposure to ambient conditions. This extends the lifespan of equipment and reduces maintenance requirements, enhancing operational reliability and cost-effectiveness.

**2. Application of Gas Line Thermal Jackets:**

Gas line thermal jackets are utilized in cleanroom environments across semiconductor manufacturing facilities, particularly in areas where sensitive processes and equipment are involved. Common applications include:

- **Gas Distribution Lines:**
Thermal jackets are installed around gas distribution lines, including bulk gas cylinders, gas panels, and distribution manifolds, to maintain gas temperatures and prevent condensation during storage, transportation, and delivery to process tools.

- **Process Gas Lines:**
Gas lines connected to process tools such as chemical vapor deposition (CVD), physical vapor deposition (PVD), etching, and ion implantation systems may require thermal jackets to ensure consistent gas supply temperatures and prevent disruptions to critical manufacturing processes.

- **Specialty Gases:**
Certain specialty gases used in semiconductor fabrication, such as silane ($SiH_4$), ammonia ($NH_3$), and nitrogen trifluoride ($NF_3$), are highly sensitive to temperature changes and prone to condensation. Thermal jackets are essential for maintaining these gases at elevated temperatures to avoid phase transitions and ensure safe and reliable delivery.

**3. Gases Requiring Thermal Jackets:**

While the need for thermal jackets depends on specific process requirements and environmental conditions, gases commonly requiring insulation in semiconductor manufacturing include:

- Silane ($SiH_4$)
- Ammonia ($NH_3$)
- Nitrogen Trifluoride ($NF_3$)
- Hydrogen ($H_2$)
- Chlorine ($Cl_2$)
- Arsine ($AsH_3$)
- Phosphine ($PH_3$)

The list provided includes common gases used in semiconductor manufacturing processes that may require thermal jackets to prevent condensation and maintain optimal temperature conditions. However, the specific gases and their requirements may vary depending on the fabrication processes and equipment used in a semiconductor facility. It's essential to consult with process engineers and specialists familiar with the specific requirements of the manufacturing environment to determine the gases that necessitate thermal insulation and heat management measures.

**Conclusion:**

Gas line thermal jackets are indispensable components of cleanroom infrastructure in semiconductor manufacturing, ensuring the consistent temperature control of process gases to prevent condensation, maintain process integrity, and protect equipment. By selecting appropriate thermal insulation materials and implementing effective thermal management strategies, semiconductor fabs can optimize process performance, enhance product quality, and minimize operational risks associated with gas handling and delivery.

## 20.20 Rapid Thermal Process RTP

Rapid Thermal Processing (RTP) is a semiconductor manufacturing technique used to modify the properties of thin films and semiconductor substrates through rapid heating and cooling cycles. This subchapter explores the principles, equipment, and applications of RTP in cleanroom environments.

**Principles of RTP:**

RTP involves the rapid heating of semiconductor wafers to high temperatures (typically between 600°C and 1200°C) for a short duration (seconds to minutes) using high-intensity lamps or lasers. This rapid heating allows for precise control over the annealing, oxidation, and diffusion processes without causing extensive thermal damage to the substrate.

**RTP Equipment:**

RTP systems consist of a processing chamber equipped with lamps or lasers for heating, as well as control systems for temperature regulation and gas flow. The chamber is often integrated into a larger semiconductor fabrication toolset, allowing for seamless integration into the manufacturing workflow.

**Applications of RTP:**

1. **Annealing:**

   RTP is commonly used for annealing processes to repair crystal lattice defects, activate dopants, or induce phase transformations in semiconductor materials.

2. **Oxidation:**
   RTP facilitates rapid thermal oxidation processes, where thin oxide layers are grown on semiconductor surfaces to create gate oxides, isolation layers, or dielectric materials.
3. **Doping:**
   RTP enables precise control over dopant diffusion and activation, crucial for creating highly controlled semiconductor device structures.
4. **Silicidation:**
   RTP is used for silicidation processes, where metal-silicon compounds are formed to create low-resistance contacts and interconnects in semiconductor devices.
5. **Nitridation:**
   RTP can be employed for nitridation processes, where silicon nitride layers are deposited or grown on semiconductor surfaces to provide insulation or passivation.

**Advantages of RTP:**

1. **Fast Processing:**
   RTP enables rapid heating and cooling cycles, reducing processing times and increasing throughput in semiconductor manufacturing.
2. **Precise Temperature Control:**
   RTP systems offer precise control over temperature profiles, allowing for the creation of uniform and reproducible thin film structures.
3. **Reduced Thermal Budget:**
   RTP minimizes the thermal budget, limiting the diffusion of dopants and reducing the risk of thermal damage to sensitive device structures.
4. **Flexibility:**
   RTP systems are highly flexible and can accommodate various wafer sizes and materials, making them suitable for a wide range of semiconductor applications.

**Conclusion:**

Rapid Thermal Processing (RTP) plays a vital role in semiconductor manufacturing, offering fast, precise, and flexible thermal processing capabilities for the fabrication of advanced semiconductor devices. In cleanroom environments, RTP systems are integral components of the semiconductor toolset, enabling the creation of high-performance integrated circuits with nanoscale features and superior electrical properties.

## 20.21 Dicing Machine

Dicing machines are essential tools in semiconductor manufacturing used to cut semiconductor wafers into individual chips or dies. This subchapter explores the operation, contamination risks, and utility requirements associated with dicing machines in cleanroom environments.

**Operation of Dicing Machines:**

Dicing machines utilize high-precision blades or lasers to separate semiconductor wafers into discrete chips. The process involves aligning the dicing blade or laser with the predefined scribe lines on the wafer surface and applying mechanical or thermal energy to cleave the wafer along these lines, creating individual dies.

**Contamination Risks:**

While dicing machines are crucial for wafer singulation, they can also pose contamination risks in semiconductor cleanrooms. Contaminants such as particles, debris, and residual fluids from the dicing process can compromise the quality and reliability of semiconductor devices. Therefore, proper maintenance, cleaning procedures, and contamination control measures are essential to mitigate these risks.

**Utility Requirements:**

Dicing machines require various utilities to support their operation and ensure optimal performance. The following utilities are commonly connected to dicing machines in semiconductor cleanrooms:

1. **Liquid Nitrogen (LN2):**
   Dicing machines often utilize liquid nitrogen as a coolant to reduce friction and heat generation during the dicing process. LN2 helps maintain the temperature of the dicing blade or laser, preventing thermal damage to the semiconductor wafer and ensuring clean, precise cuts.
2. **Deionized (DI) Water:**
   DI water is used for cleaning and rinsing semiconductor wafers before and after the dicing process. It helps remove particles, residues, and contaminants from the wafer surface, ensuring the quality and integrity of the diced dies.
3. **Sewage System:**

Dicing machines generate waste materials such as debris, slurry, and cutting fluids during the dicing process. A sewage system is connected to the dicing machine to collect and dispose of these waste materials safely and efficiently, preventing environmental contamination and maintaining cleanroom hygiene.

**Conclusion:**

Dicing machines play a critical role in semiconductor manufacturing by enabling the precise singulation of semiconductor wafers into individual chips. However, they also present contamination risks that must be addressed through proper maintenance, cleaning, and contamination control practices. By ensuring the availability of essential utilities such as liquid nitrogen, deionized water, and a sewage system, semiconductor cleanrooms can optimize the performance of dicing machines while maintaining strict cleanliness standards and minimizing the risk of wafer contamination.

## 20.22 Laser Excimer

Laser Excimer technology plays a pivotal role in various processes within semiconductor manufacturing. This subchapter elucidates the operational principles, applications, and significance of laser excimer systems in semiconductor fabrication.

**Operational Principles:** Laser Excimer systems utilize a gas discharge laser to emit short pulses of ultraviolet (UV) light with wavelengths typically in the range of 193 to 248 nanometers. The excimer laser's operation involves the use of a noble gas, such as xenon or krypton, and a halogen gas, such as fluorine or chlorine, in a sealed chamber. When an electrical discharge is applied to the gas mixture, it undergoes a rapid transition from an excited state to a ground state, emitting UV light in the process.

**Applications in Semiconductor Manufacturing:**

1. **Photolithography:**

   Excimer lasers are extensively used in photolithography processes for semiconductor patterning. The UV light emitted by excimer lasers is employed to expose photoresist-coated semiconductor wafers, facilitating the transfer of intricate circuit patterns onto the wafer surface with high precision and resolution.

2. **Annealing:**

   Excimer laser annealing is employed to modify the material properties of semiconductor films and substrates. By delivering short pulses of intense UV radiation, excimer lasers can induce localized heating and crystallization in semiconductor materials, enhancing their electrical conductivity and performance.

3. **Ablation:**
   Excimer lasers are utilized for material removal processes, such as laser ablation and etching, in semiconductor fabrication. The high-energy UV pulses emitted by excimer lasers can selectively remove or modify thin films, dielectric layers, and other materials on semiconductor substrates with minimal damage to surrounding structures.

4. **Doping and Ion Implantation:**
   Excimer lasers are employed in ion implantation processes to activate dopant atoms and modify the electrical properties of semiconductor materials. The UV radiation generated by excimer lasers can facilitate the diffusion of dopant ions into semiconductor substrates, enabling precise control over dopant concentration and distribution.

**Significance in Semiconductor Manufacturing:**

1. **High Precision and Resolution:**
   Excimer lasers offer unparalleled precision and resolution in semiconductor patterning and material modification processes, enabling the fabrication of advanced semiconductor devices with nanoscale features and dimensions.

2. **Versatility:**
   Excimer laser systems can be adapted to various semiconductor manufacturing processes, including photolithography, annealing, ablation, and ion implantation, making them indispensable tools for semiconductor fabrication facilities.

3. **Process Control and Optimization:**
   Excimer laser technology enables precise control over critical parameters such as energy density, pulse duration, and wavelength, allowing semiconductor manufacturers to optimize process conditions and achieve desired device characteristics with high repeatability and reliability.

**Conclusion:**

Laser Excimer technology plays a crucial role in semiconductor manufacturing by enabling precise patterning, material modification, and process control in various fabrication processes. With their versatility, high precision, and reliability, excimer laser systems contribute significantly to the advancement of semiconductor technology and the production of state-of-the-art semiconductor devices.

# Appendix A: Cleanroom Glossary

1. **Chemical Storage Cabinets with Exhaust:**
Safely stores and handles chemicals within the cleanroom environment while ensuring proper ventilation to mitigate potential hazards.

2. **Critical Drier:**
A specialized drying apparatus used to remove moisture from wafers after wet processes, ensuring optimal surface conditions for subsequent fabrication steps.

3. **DI Water Station:**
Provides deionized water, free from impurities, for various cleaning and processing steps within the cleanroom environment.

4. **Dicing Machine:**
Separates individual MEMS (Micro-Electro-Mechanical Systems) devices from the semiconductor wafer, enabling precise cutting and packaging of microelectronics components.

5. **Ellipsometer:**
Measures the thickness and optical properties of thin films deposited on semiconductor substrates, providing critical data for process control and film characterization.

6. **Exhaust Systems**:
Ventilation systems designed to remove contaminants, fumes, and airborne particles from cleanroom environments, ensuring compliance with cleanliness standards and worker safety regulations.

7. **Furnaces Types:**
Various types of furnaces used in semiconductor processing, including tube furnaces, rapid thermal processing (RTP) systems, and atmospheric-pressure chemical vapor deposition (APCVD) reactors, each tailored for specific thermal processes such as oxidation, annealing, and deposition.

8. **Gas Cabinets:**
Stores and supplies gases required for semiconductor fabrication processes, maintaining a controlled and safe environment for handling hazardous materials.

9. **Gas Scrubbers**:
Safely disposes of exhaust gases generated during semiconductor processing, minimizing environmental impact and ensuring compliance with safety regulations.